50th Anniversary of the SE Missouri Ozarks UFO Flap:

Piedmont, Clearwater Lake, Farmington, Wayne County

Debbie Ziegelmeyer

and

Margie Kay

50th Anniversary of the SE Missouri Ozarks UFO Flap:

Piedmont, Clearwater Lake, Farmington, Wayne County

By Debbie Ziegelmeyer and Margie Kay

ISBN: 978-1-7378996-6-2

Printed in the USA

Publisher:
Un-X Media
PO Box 1166
Independence, Missouri 64050
www.unxmedia.com

Testimonials

Topnotch UFO investigators Debbie Ziegelmeyer and Margie Kay have written a fascinating page-turner on recent and historic UFO sightings and ET contact in Missouri. You will read the testimony of credible witnesses, view the evidence, both physical and physiological, and feel astonished by the presence of highly advanced aerial vehicles and their mysterious occupants.

Kathleen Marden
Author of *"Extraterrestrial Contact: what to do when you've been abducted"*

The 1973 UFO Flap in Piedmont and Wayne County Missouri has long been forgotten by many UFO investigators and enthusiasts, but this book brings the event to the forefront once again. Since UFO sightings have been ongoing in this part of Missouri for many years it is worth keeping an eye on this area.

Craig Campobasso
Author of "The Ufo Hotspot Compendium"

In a most fitting tribute to the "Show Me" state of Missouri, top UFO investigators Margie Kay and Debbie Zieglemeyer bring together an impressive collection of UFO reports, credible witness testimonies and presented evidence from past to present in their compelling new book ..."50th Anniversary of the SE Missouri Ozarks UFO Flap". It's chocked full of insightful information about the "Show Me" state of little known but very active UFO history over the last 50 years plus updated interviews. You will read about reports of advanced aerial craft with mysterious beings aboard. Fascinating newspaper clippings of the Ozarks area UFO sightings from the 70s reveal surprise glimpses into a different time in history and the media from our world today from David Marler at the National Historical UFO Records Center. Authors Margie and Debbie have something for everyone at the end with segments including "What to do if You See a UFO/UAP" ..."About Missouri MUFON and How to File a UFO Report"... along with a list of social media UFO/ET groups to join if desired. What an astonishing treasure trove of Missouri's UFO/ET history mysteries!

Cheryll Jones
Coast to Coast AM Investigative Reporter
Independent Journalist, former CNN News Anchor

Acknowledgements

I want to thank Dave Marler for not only encouraging me to write this book, but for providing me with numerous copies of 1973 newspaper articles from the Brushy Creek area.

Debbie Ziegelmeyer

Our appreciation and greatest respect to The Mutual UFO Network and its volunteers who have spent months and in some cases years, investigating sighting reports of unknown lights and craft all at their own expense. This organization has investigated and documented sightings for decades archiving their findings for those such as myself who wish to collect and use this data as proof of the existence of other worldly beings and their visitation of our planet Earth.

Dedication

This book is dedicated to the late Harley Rutledge, PhD, who devoted years of his life to the study of the unexplained phenomena in southeast Missouri, and to the late Ted Phillips, who dedicated much of his life to trace evidence study, and the study of strange phenomena in the State of Missouri.

Contents

Forward

By Lee Speigel
Host of the Edge of Reality radio show and journalist

When it comes to anything to do with the topic of UFOs, the most important question that always comes up is: What's the evidence for this or that particular UFO case? In Missouri, a higher need exists for issues that require strong evidence: hence, the nickname often given to the "Show Me State."

Missouri hast often been in the grips of what is commonly referred to as a "UFO Flap" – very large numbers of unidentified flying objects reported by citizens. The 1970's UFO Flap is centered around the Piedmont, Clearwater Lake, Farmington and Wayne County areas in the southeast of the state, where it's the 50th anniversary of this particular series of unexplained events that began in 1973.

Over a period of just a few months that year, more than 500 UFO reports were reported to local law enforcement officials. The ongoing investigation of this "flap" is in the capable hands of two seasoned and respected UFO researchers and authors.

Debbie Ziegelmeyer is State Director of Missouri MUFON (Mutual UFO Network – the world's oldest and largest UFO organization), serves on the MUFON board of directors and is Director of the Underwater Research and Recovery Team. She's also author of "The Alien Colonization of Earth's Waterways: A Reference Guide to UFO/USO Water Related Activity."

Margie Kay brings 40 years experience as an investigator and author to uncovering the facts of paranormal activity throughout the United States. She's the owner and CEO of the UnX Network and Un-X Media publishing company, serves as Assistant State Director for Missouri MUFON and is Director of the OZ Interdimensional Communication Institute. Margie is the author of several books, including "The Kansas City UFO Flaps," "The Fast Movers: Evidence of High-Speed UFOs," "Haunted Independence Missouri," "Gateway to the Dead: A Ghost Hunter's Field Guide," "The Remote Viewing Workbook" and "A Sonoma County Phenomenon." She's also host of the "Un-X News Radio/Podcast" on her YouTube Channel.

With a combined total of nearly 3,000 UFO and paranormal investigations between them, Debbie and Margie set out to examine and share the 50-year-long Missouri UFO reports that continue to this day.

Missouri has the distinction of being the only other state (besides Tennessee) that is surrounded or touched by eight separate states. It's also known by several nicknames, including the "Show Me State." One theory of the origin of that phrase suggests that Missourians aren't gullible and not willing to believe something without proper evidence. That attitude certainly holds true to any discussion about UFOs.

Debbie Ziegelmeyer and Margie Kay's new book, "50th Anniversary of the SE Missouri Ozarks UFO Flap," chronicles the many reported sightings of unexplained craft in the Piedmont, Missouri, region in 1973.

In October, 2011 one of the biggest UFO flaps occurred in which 97 reports were filed in the greater Kansas City area. In July of 2019 Three white object appeared over Kansas City that remain unexplained. The entire city was in panic, with no answers coming from the Media. In August through October of 2022, UFO sightings in Missouri were reported at three times the normal rate, and a month earlier, massive, grey metallic-colored objects were spotted in six locations in the northwest part of Missouri. Eyewitnesses described seeing two rows of windows, lit from within the side of the craft. No exterior lights were reported on the silent objects.

Other UFO reports out of Missouri have included sightings of round and triangular objects. But it was the sighting of a large unidentified craft hovering over a field above treetops on February 21, 1973, that ushered in the UFO flap. The eyewitnesses included a basketball coach and five members of his team who were en route home from a tournament in Essex, Mo. When several other eyewitnesses corroborated this story with their own descriptions of unexplained lights "dancing in the sky" as well as large, mysterious craft hovering over fields, this had a huge impact on the area that brought reporters, scientists and, of course, other citizens who just wanted to see these unexplained objects.

Among the scientists who visited Missouri to investigate those UFO reports was Dr. J. Allen Hynek, the Chairman of Northwestern University's Astronomy Department and who had previously been a longtime consultant to the United States Air Force's infamous Project Blue Book UFO study. Also with Hynek was Ted Phillips of Sedalia, Mo., a MUFON State Section Director and a recognized UFO landing traces expert in the field. The fact that Hynek took an interest in these Missouri sightings was a strong indicator that he – a former UFO skeptic – wasn't disregarding the many reports emerging from this region.

Five years after these initial Missouri sightings, Hynek joined this writer and several other notable scientific and military individuals to present evidence of UFO reality to the United Nations. In fact, Hynek very boldly told the UN delegates exactly how important UFOs were (and still are) to the world.

"There exists today a world-wide phenomenon," Hynek said. "Indeed, if it were not world-wide, I should not be addressing you and these representatives from many parts of the world. It is a phenomenon so strange and foreign to our daily terrestrial mode of thought that it is frequently met by ridicule and derision by persons and organizations unacquainted with the facts. Yet, the phenomenon persists; it has not faded away as many of us expected it would when, years ago, we regarded it as a passing fad or whimsey. Instead, it has touched on the lives of an increasing number of people around the world.

"In the popular mind, the UFO phenomenon is associated with the concept of extra-terrestrial intelligence, and this might yet prove to be correct in some context. Surely it would be a huge joke on all of us should it develop that evidence for extra-terrestrial intelligence was right here under our noses all the time while great effort was spent searching for it elsewhere! The UFO phenomenon, as studied by my colleagues and myself, bespeaks the action of some form of intelligence. But whence this intelligence springs, whether it is truly extra-terrestrial, or bespeaks a higher reality not yet recognized by science, or even if it be in some way or another a strange psychic manifestation of our own intelligence, is much the question."

"50th Anniversary of the SE Missouri Ozarks UFO Flap" is a long-awaited tribute to all the people who broke through the ridicule barrier that infested the UFO topic for many decades. Those individuals found the courage to speak up and tell others what it was they saw in the sky that they couldn't understand or who searched for like-minded people who also wanted to share their experiences.

As you get deeper into the book, it becomes very apparent that a lot of people – not only in Missouri, but in many other states and countries – are experiencing something similar that science has yet to organize into a coherent and globally accepted answer to the UFO enigma.

This book is also a big shout out to Debbie Ziegelmeyer, Margie Kay and the untold numbers of dedicated researchers and writers and caring individuals who put their reputations on the line in the age-old quest for ultimate answers to the UFO subject which will likely have a profound impact on all citizens of Earth. Eyewitnesses no longer need to feel embarrassed after having a UFO experience and keeping it to themselves out of fear of ridicule.

Kudos to Debbie and Margie for their energy and intelligence as they continue to open the doors to UFO reality and acceptance – doors that were once closed to the public. And their work shouldn't be too surprising, after all, in a place called the Show Me State.

Introduction

What is known in ufology as a "UFO Flap" is an unusually large number of unidentified craft sightings reported by the public in a short span of time. Most believe the 1973 UFO Flap to be confined to the Piedmont, Missouri area, but similar UFO Flaps were occurring in other states during the same time period. Beginning in February 1973, the Piedmont, Missouri police department received over five hundred calls from area residents who were reporting UFO sightings in the Brushy Creek area. Brushy Creek, which is in southeastern Missouri, borders Piedmont to the north and the town of Mill Springs nine miles to the south. The area is best known by Missouri residents for its two recreational lakes which are Clearwater Lake just seven miles west of Piedmont, and Wappapello Lake forty miles southeast.

Although sightings in the Piedmont area began years earlier, February 21, 1973, would prove to be the beginning of continuous witness reports so numerous, national media would make their way to this small town in troves. The first reported UFO sighting on the evening of February 21, 1973, was by Reggie Bone and five members of his basketball team who were on their way home from a basketball tournament in Essex. This was followed by two other noteworthy sightings, one the same night by Edith Boatwright, and a third sighting one month later, on March 21, 1973, over Clearwater Lake by Jean Coleman and Kathy Leach.

The sightings continue to this day, according to residents who were interviewed, and their stories are included in this book as well. It seems that the UFOs never left southeast Missouri.

Piedmont Missouri

While investigating an area where a "UFO Flap" has occurred, it is important to research its history as far back as feasible. An understanding of the land and its people could lead to why there is an extraterrestrial presence. Something in the Brushy Creek area is and has been, attracting life from far-off worlds for decades if not longer. To look at Piedmont and the area, it would seem at least on the surface, to be a normal midwestern town. However, just west of the Mississippi river between the Missouri and Arkansas River in lands controlled by the Osage Indians, an ancient meeting was held between settlers, native Americans and "star people".

As told in their stories by the Osage Indians, they recalled a "great mystery force." The grandfather of the Sun, Wah-Kon-Tah, decided that these people of the earth were dirty, confused, and lived in chaos. He sent a number of star people who floated to the earth with their arms outstretched like wings, in hopes of teaching the earth people the order of the universe and how to live in harmony. As legend has it, the earth welcomed the star people by dropping a shower of acorns which rattled the leaves of the oak trees as they fell from the sky. Thus, a friendship was forged between the star people and the earth people who agreed to adopt the orderliness of the universe as taught by the star people.

The formation of the town of Piedmont, Missouri did not begin with the name "Piedmont" which was renamed in later years. Two brothers, James and William Daniel were the first people to hold official land entries ownership from 1853 to 1856 for the Piedmont area. In the beginning, the town was known as Danielsville after the brothers, rather than Piedmont. The brothers who had just fought together in the Spanish - American War were previously farmers. They purchase a large amount of property in the area as a means of getting back to normal life which for them, was farming. James Daniel was lucky, his property was suitable for farming but unfortunately, his brother William's property was not. William's property ran along the eastern side of McKenzie Creek so most of it was very marshy with hills. Because of this, brother William, decided to open a general store which was much needed in the area. This store, and the fact that land in the area was unbelievably cheap, attracted many settlers. This onslaught of residents helped to establish the village of Danielsville.

Before long, enough families with children had settled in the area which brought the need for a school. The Blue Moon Schoolhouse was established which also served as a place to hold church services for the area. The village began to grow very quickly, and in 1855, a post office and a stagecoach service were established. With the stagecoach service running from Ironton to Doniphan, Danielsville's population continued to grow reaching almost three hundred residents in the first four years.

In 1868 a decision was made in New York to extend the railroad track of the St. Louis Iron Mountain and Southern railways all the way to the Arkansas state line. Trains with work crews were operating as far south as Piedmont beginning as early as 1870. Crews living in Danielsville were working long hours to extend the railway to the Arkansas state line. This was a requirement before any regular scheduled service could begin in the area. Finally, on October 10, 1871, the first regularly scheduled operated train made its way from St. Louis to Poplar Bluff.

Earlier in 1871, Thomas Allen, the Vice President of the railroad company, traveled to Danielsville to make a real-estate land grabbing deal to monopolize land and properties before the arrival of the train. The property Thomas Allen wanted to purchase belonged to James Daniel who thought the offer was too risky. Later that year Thomas Allen was able to work out a deal with James' brother William Daniel who purchased the property from his brother and then held the note on the property. Thomas Allen divided up the property into business lots which sold at a price that was profitable to both Thomas Allen and William Daniels. The railroad rush of 1871 had started, the local economy boomed with railroad workers, farmers and new businesses all flocking to the area. Later in 1871, when the railroad was extended to Danielsville, the town was renamed Piedmont by Thomas Allen which is French (pied and mont) meaning "foot" and "mountain." The name of the post office was also changed to Piedmont in 1872.

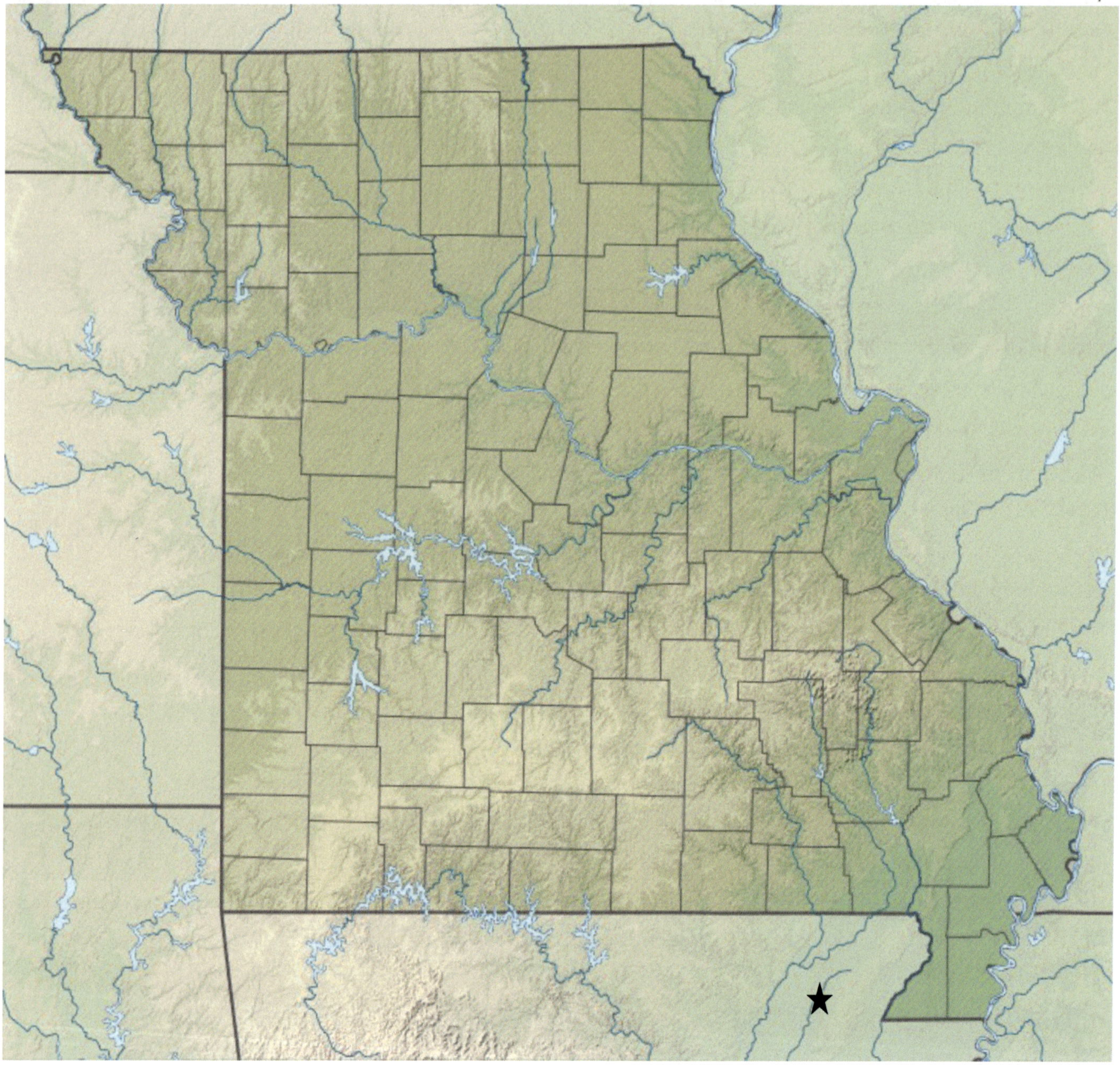

Topographical map of Missouri showing the location of Clearwater Lake

Photo credit: SANtosito, CC BY-SA 4.0 <https://creativecommons.org/licenses/by-sa/4.0>, via Wikimedia Commons

Clearwater Lake is a reservoir on the Black River, six miles (9.7 km) from Piedmont, Missouri. The U.S. Army Corps of Engineers uses Clearwater for flood control in the White and lower Mississippi River Basins.

Clearwater Lake was so named on account of its clear, spring-fed water. Construction began in 1940 but was halted temporarily at the advent of World War II. **Clearwater Lake Dam** opened in 1948 as an earthen and concrete dam, 114 feet (35 m) high. The lake has a surface area of about 2.5 sq mi (6.5 km^2). Source: Wikipedia

Chapter 1

Historic Brushy Creek

By Debbie Ziegelmeyer

Clearwater Lake

Clearwater Lake is a reservoir on the Black River six miles from the town of Piedmont, Missouri in Reynolds County. The original purpose of the lake was for flood control in the White and lower Mississippi River Basins. But today it has become one of the largest tourist attractions throughout Southeast Missouri. The Little Rock Corps of Engineers estimates that over a million people per year travel to Clearwater Lake for camping, canoeing, boating, fishing, and swimming.

The early history of the land where Clearwater Lake is now located, goes back to the Paleo-Indians. The Paleo-Indians camped and hunted along many of the Ozark Rivers as far back as 14,000 years ago. The Osage Indians were in this area around the time of the American Revolution, and the Shawnees and Delawares also in this area, were the first Indians the English-speaking settlers found. Therefore, many of the local place names, roads and attractions are based on the Shawnee.

The construction of Clearwater Lake Dam began in 1940 but was put to a temporary halt with the start of World War II. On Memorial Day, May 30, 1948, the Clearwater Lake Dam was finally officially opened and witnessed by various state and federal officials including top officers in the Corps of Engineers Little Rock District. The event was also attended by Lieutenant Governor James T. Blair, Jr. The lake dam opened as an earthen and concrete dam; 114 feet high with a lake surface area of about 2.5 square miles. 5,500,000 cubic yards of earth had been used to make the dam with an elevation to the top of 608 feet. The dam's emergency spillway was 190 feet long with a crest of 567 feet and a spillway tunnel length of 1,771 feet long with a diameter of twenty-three feet. But, due to work being shut down during World War II, the entire project was not fully completed until September 1958.

Currently, Clearwater Lake has 1,630 acres at the top of the conservation pool during the winter months, a flood pool top of 10,350 surface acres and a total storage capacity of over 413,000 acres feet of water. It is ranked number three on the list of Missouri recreational lakes.

Native American Presence

Missouri Native Americans occupied the Black River Recreation Area of Missouri going back to the Paleo-Indians. These ancient Americas who were present at the end of the last ice age, camped and hunted along Ozark waterways as far back as 12,000 to 14,000 years ago. They were big-game hunters of mostly the mastodon for meat and the giant ground sloth for fur. Around 8000 B.C. when the ice age arrived causing the disappearance of the large mammals, these native Americans began to hunt smaller game, fish, and use vegetables as a new food resource.

The Hopewell tribe moved into the region now known as Missouri during the "Woodland Period" of 1,000 to 500 B.C. They cultivated corn, hunted deer and wild turkey, and learned how to fire clay pots and tools to trade with other tribes. During the "Mississippian Period" from 900 to 1700 A.D., the Native Americans became even more dependent on the rivers for food and growing crops. The last classified era of Native American development was the "Historic Period" which began in 1700 and was the period where European explorers and settlers came into contact with the various tribes. This area was the hunting grounds for many tribes including the Delaware, Kickapoo, Piankasaw, Osage, and Shawnee.

The first recorded contact with a native American tribe was in 1673 when French explorers Louis Joliet and Jacques Marquette crossed paths with the Osage tribe who dominated the area. This was the only truly native tribe to the Ozark Missouri region, the rest being tribes driven west towards the Mississippi River by the growing presence of the white man. From the book "History of Early Reynolds County, Missouri, by James Bell, the Osage Indians lifestyle was described as follows; "Due to their marriage customs, the Osage were tall, physically strong, and possessed unquestionable courage. The smaller, weaker males often were denied marriage and the mightiest warriors got the girl plus all her sisters. In this way they had a form of selective breeding, which undoubtedly accounts for most of the tribe being over six feet tall." Their empire covered a massive land span which included what is known today to be Missouri, Kansas, Oklahoma, and Arkansas with an estimated population of over 20,000. In a treaty with the Federal government drafted in 1808, the Osage Indians relinquished most of their claim to the Ozark Plateau which was later described as one of the areas where the "Trail of Tears" passed through. Several native American tribes hunted and camped in the area known today as The Brushy Creek area and considered it sacred grounds. Part of this sacred area was dammed and flooded in the early 1940s to form Clearwater Lake.

Photos of Clearwater Lake

Credit: Debbie Ziegelmeyer

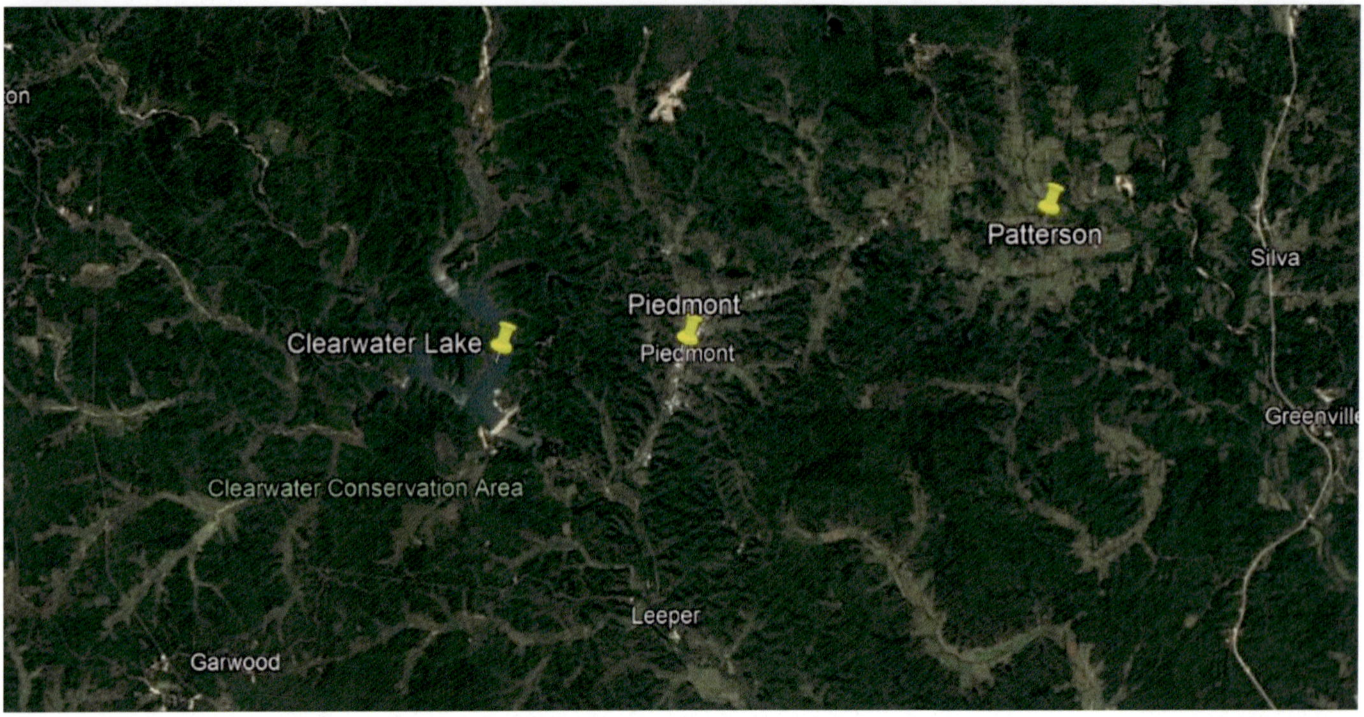

Location of Piedmont , Clearwater Lake, and Patterson   Credit: Google Maps

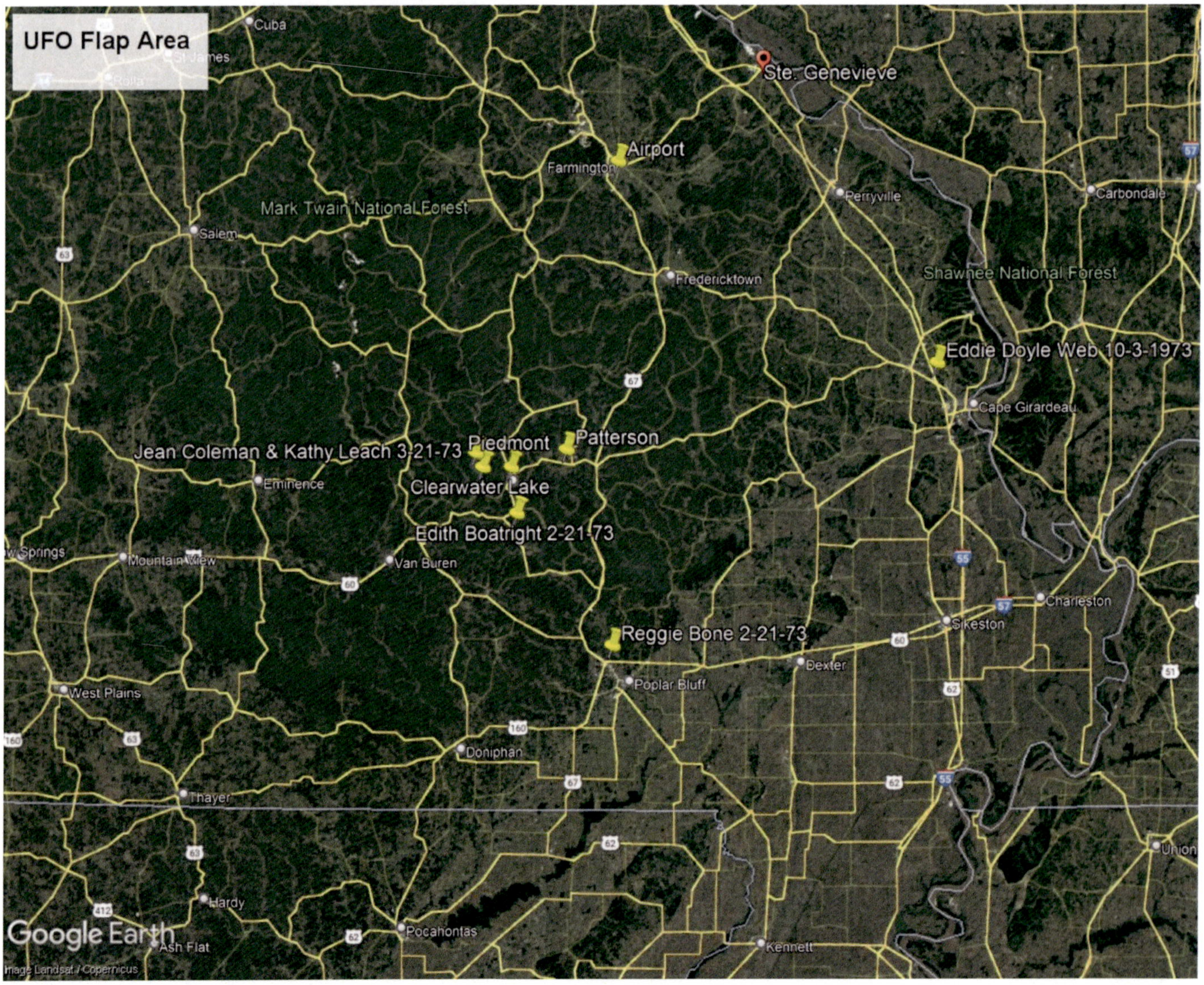

Chapter 2

Most Notable Sightings

By Debbie Ziegelmeyer

Reggie Bone
February 21, 1973, 8:30pm

Basketball coach Reggie Bone and five members of his basketball team were returning from a tournament in Essex, Missouri when they came upon a large unidentified craft hovering in a field just above treetop height. Numerous additional witnesses came forward with descriptions of unexplained lights "dancing in the sky" and large unknown aircraft hovering over fields. This sighting triggered a massive event in the area bringing reporters, scientists, and spectators hoping to see something unexplained.

The following is part of a transcribed interview by Edward O'Herin, New Madrid, Missouri, MUFON State Section Director for southeast Missouri. The interview which took place on March 24, 1973, is of Reggie Bone's sighting of Thursday February 21, 1973, 8:30pm

What is your name?
Reggie Bone

And you live in Piedmont?
Right

About how old are you Mr. Bone?
43

What is your occupation?
Teacher, basketball coach

Would you tell me what you saw with reference to these objects or lights around this area recently?
Well, February 21 we were coming back from a basketball tournament at Essex and as we approached the turn off on 60 which is highway A, it all started. I was kidding one of the kids, the one who keeps score for me, because he missed two points on our score book someway and I just happened to look over and these lights were in the tree line and I said well, Fred, you can see those lights up there can't you and the rest of the kids in the car got to looking and they said yes. I looked back, we just passed it off

Newspaper Clippings:

Photo by Walt Andrus

GLOOMY DRIVE HOME FROM A BASKETBALL DEFEAT turned into an adventure for Clearwater High School basketball player Cary Barks (foreground) and coach Reggie Bone (at right), who claim they watched a UFO for 10 minutes.

The photographer, Walt Andrus, was the Director of MUFON

Hayden Hewes of Oklahoma City's International UFO Bureau interviews basketball coach Reggie Bone while Daniel Garcia of IUFOB takes notes.

VOL. 95 NO. 81 © 1973, St. Louis Post-Dispatch

Reggie Bone

The basketball coach who saw the UFO after a losing game

as an airplane. I looked back and I thought it sure doesn't have the markings of an aircraft, but I didn't say anything to the kids and just thought to myself that thing hadn't moved a lick.

And, then we turned west on 49 and come across the Brushy Creek area and there is also an area down there called James Hill and go up over a hill and there is a church on top of the hill and as you come off this hill one of the kids in the back seat said; "Coach there is what we saw over on 60 sitting right out there in the field!"

And of course, we got the car screeched to a stop and we all got out and slammed car doors and made a mad run to the back of the car to see it, and what we saw were just a band of lights. I couldn't say whether they were actually revolving like your wheel there, or whether they were blinking on and off but there was no pattern. We looked for a pattern and discussed it later. There was red, green, amber, white, you might see red, green, then red again, then amber and then white. But we estimated this to be fifty feet off the ground and in that valley and there was no moon on this night. It was close enough; I would estimate two hundred yards and that is being conservative, really.

I still thought at this time that anytime now we will hear something you know motor, or it could be a copter down and in trouble or something of this nature. But then, suddenly, the thing began to ascend straight up. We could still see the lights. It was real dark, and the lights were very distinct. I thought oh, oh and the kids, there was

no talk among the kids, nobody said a word. I couldn't estimate at what height that it reached before it stopped again. Only the fact that it did not get above the tree line.

What was between you and this object?

There is a huge field, there is a large hill that will set north of the field, and the Black River is north of the place where we were, and the rest of it is I would say is farmland and some swamp land and some timber.

After you stopped there, did you turn off your lights and motor?

We did turn off our lights yes, and we turned off the motor.

What did it do then?

Well, it made a 90 degree turn, went over in the tree line to our right. It got over there and stopped, we watched it and it began to circle back around the field over closer to the tracks and the river and I told the boys, I said, let's get in the car and go and there wasn't any of them raised any objection. We got in the car, and nobody said a word and we continued on and all of a sudden one of the kids said to me; "Coach that thing is back of the car back here!" I can't truthfully say that it was following along the car, but it had gotten lower, and it was in back of the car and at this time I would say about a fourth of a mile.

I didn't want to stop driving. I thought well, there is a little town up here of Mill Springs, it does have streetlights and maybe we can get some outline of the object if we can get up there in town and the object is still close. We went across Black River bridge which was right along in this vicinity that the object turned and never did get above the tree line and disappeared behind Spring Hill.

Did it ever increase its speed?

It never seemed in any hurry.

You never heard anything?

There was no noise even on its upward rise.

Was there any flashing?

It was hard to determine really if they were revolving or flashing, but there was no pattern.

What would you say that the distance was between any of the lights?

I asked the boys with me and of course, we got five different answers. To me, I would say they were somewhere between four and five feet apart.

Were they all in the same plane?

None of them were higher than others.

No beam. The thing didn't ever seem to tilt or anything. All the movements we saw were precise movements. I mean it didn't seem to dart. Like I said, it moved about very freely and moved like it didn't seem to be in any hurry.

The object never made it to the lighted area, so the coach and the five basketball players were not able to make out an outline. The car radio was on, and they had been listening to a ball game without experiencing any radio interference.

The next day after the incident, Reggie Bone met with the boys at school to see how they were doing. They all agreed that their best description of what they saw was "weird." "Very Weird."

All of the boys involved live in town and two or three of them work Saturdays at the IGA grocery store.

Reggie Bone asked one of the boys from school Sam Willard, who lives near to the area, to go back during daylight and take a look around. Sam walked the field but found no sign anything had been there. There was nothing to scorch that time of year and no marks were found. Sam's mother Mrs. Willard told her son she had seen something in the field that night. She ran into the house to get the other family members but when they went outside the object was gone.

Witnesses:

Clearwater High School Basketball Coach Reggie Bone

Players:

Tom Martin, freshman, age 15

Fred Allen, freshman, age 15

Randy Holmes, junior, age 16

Greg Twidwell, junior, age 16

Cary Barks, junior, age 16

Reggie Bone goes on to tell of another sighting he had the night before on March 23, 1973.

Reggie Bone was sitting outside on the top of a hill waiting for his friend Bill, to pick him up. He was looking in the direction of Ross Motor Company when he noticed a great big amber light. At first, he thought it was a streetlight from Ross Motors but when he looked to his left, he confirmed that light. So, what was he looking at? Suddenly the amber light began to glow. Reggie went into the house to get his wife who was on the phone. She dropped the phone and ran outside but in the wrong direction. When Reggie's friend Bill arrived, Reggie pointed out the object and asked Bill what he thought? Bill yelled one of his famous lines "mercy, mercy, mercy", yelled at him to get into the truck and go catch the thing. Reggie said he told Bill they should stay put because others who had seen and de-

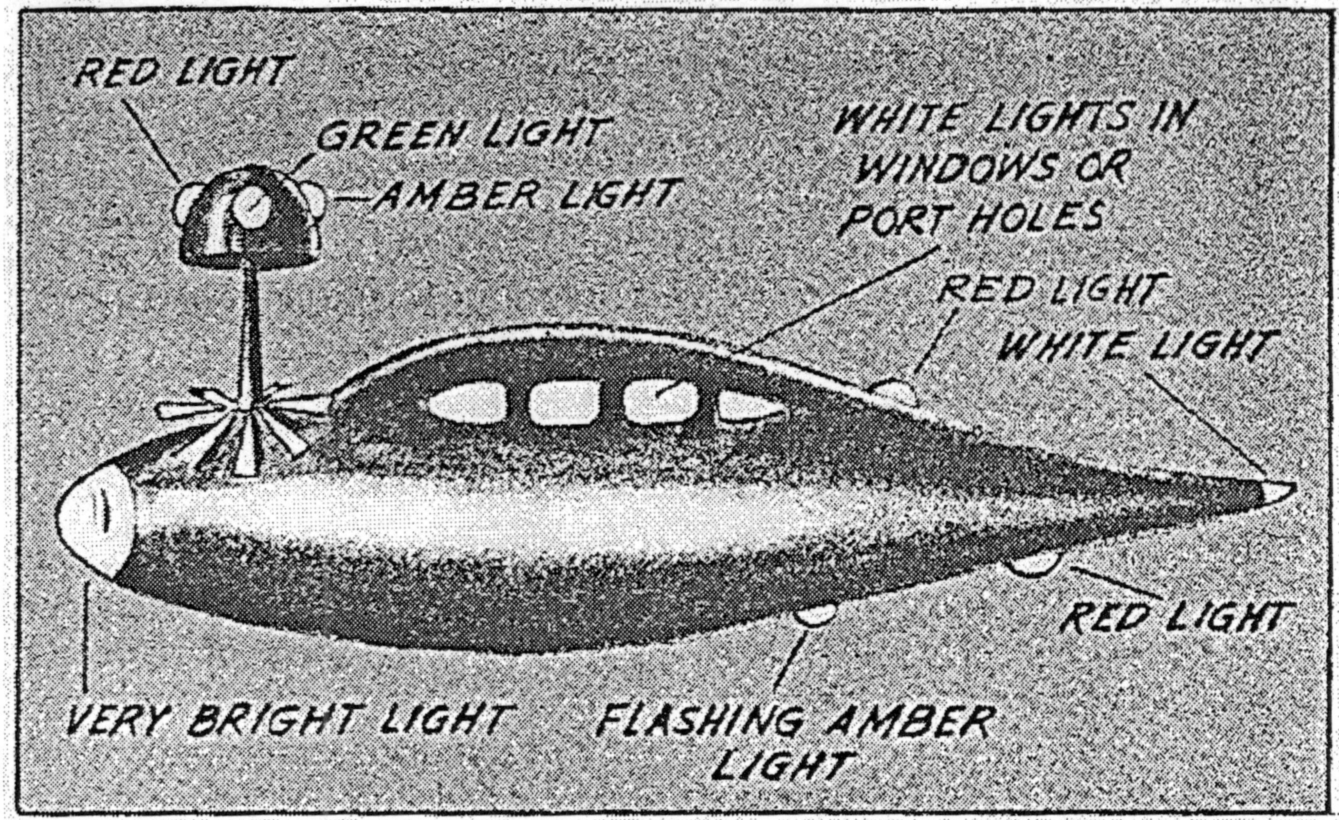

Edith Boatwright of Mill Spring, Mo., sketched the UFO she observed near her farmhouse on the night of February 21, 1973, as it flew low with its lights flashing.

scribed these lights and say they illuminate to bright white just before they are ready to move, and they will get to see this. Bill believed the object was up Oasis Road so off the two went but unfortunately one hill too far. When they arrived, they could still see the light which started jumping around the sky and then moved towards Brushy Creek.

Pat B. also witnessed the same light that night. He believed the time to be 7-7:30pm because he first saw it right after watching "Sanford and Son" on TV. He described the light like Bill had, as jumping in the sky while moving away towards Brushy Creek.

Edith Boatwright

February 21, 1973, 10:00pm

Approximately thirty minutes after the sighting by Reggie Bone and his player, another report came in at 10:00pm from a woman in Mill Spring named Edith Boatwright. Mrs. Boatwright was lying on her bed almost asleep when she noticed flashing lights outside reflecting in through her partially pulled curtains. Since her home is near the highway, Mrs. Boatwright thought the lights to be a traffic accident. As she gazed over her bedroom curtain, she noticed a strange craft flying low just barely clearing the utility lines. The craft, which was in a horizontal position, was emitting a very quiet "whooshing" sound. The craft moved slowly and evenly at a low altitude just below the tops of the trees. When the craft changed to a vertical position the emitted sound grew louder and the sound seemed more like a motor "pulling." She described the craft as long with the shape of a helicopter with no chopper blades on top. The front of the craft had rotary blades and an umbrella-like part that extended upwards. It was approximately thirty feet long with a beautiful light-colored fuselage and a darker tail. The craft was just 200 to 250 yards from her bedroom window just below the tops of the nearby Oak trees. Mrs. Boatwright watched the craft for one to two minutes before she lost sight of it from her vantage point.

Jean Coleman and Kathy Leach
March 21, 1973, 8:00pm

The most famous of the Clearwater Lake sightings was that of Jean Coleman and Kathy Leach.
Jean drove to the city dump because she had heard some people were there looking for this strange light. It was a little past 8:00pm on March 21, 1973, when she met up with Kathy Leach and the two decided to drive to Clearwater Lake. As the ladies were crossing the Clearwater Lake dam, they noticed strange lights near the Bluff View boat dock. As a car approached them which they later found out was Ken, they noticed a red flash out over the lake. Jean and Kathy got out of the car in time to see the red flash ascend. Jean got back into her running car because she had both of her daughters with her but kept the windows down. Kathy remained outside of the car for a better view. Suddenly, five lights arranged horizontally had come on right over the water not too far from the surface. The large red light was no longer visible. They watched the lights turn off

Debbie Ziegelmeyer (left) and Jean Coleman (right)
photo by Debbie Ziegelmeyer

and back on two or three times in short quick flashes then stayed on as the craft circled the lake as if it were going to fly over Piedmont. When
21, 1973, the same night Reggie Bone and his basketball players had their sighting. She and her husband were returning from a Rotary Club meeting which they had left at 8:00pm. At approximately 9:00pm, they saw a white flashing light bouncing around in the sky. Jean's husband believed the light bounced up at least fifty feet with each bounce, paused, and then would bounce again. Finally, the light flew a straight path that took it over a hill and out of sight. They did not see the light again that night.

Jean and Kathy first noticed the craft, it was below

the level of the dam, with its highest altitude being only about 20 degrees above the horizon. The duration of the sighting there at the dam was approximately three to four minutes. The ladies decided to head back to the dump to see if the craft was visible from that vantage point.

When they arrived at the dump Jean's brother was there. She asked him if he had seen anything and he replied, "Yes, it is going over right now!" Jean and Kathy got back out of the car in time to witness a large yellow, amber colored light which was not flashing, moving across the sky heading in the direction of Brushy Creek. Jean and Kathy were not sure if this light was the same craft they had seen earlier at Clearwater Lake.

Jean Coleman's first area sighting was on February 21, 1973, the same night as Reggie Bone and his basketball players. She and her husband were returning from a Rotary Club meeting which they had left at 8:00pm. At approximately 9:00pm, they saw a white flashing light bouncing around in the sky. Jean's husband believed the light bounced up at least fifty feet with each bounce, paused, and then would bounce again. Finally, the light flew a straight path that took it over a hill and out of sight. They did not see the light again that night.

Jean Coleman was a high school teacher at the time of this incident back in 1973, but now owns a clothing store in Piedmont. On March 7, 2016, Lynne Mann, Missouri MUFON area SSD and I had the pleasure of meeting and speaking to Jean. She is a very lovely woman who welcomed us into her store and greeted us with kindness. Jean had no problem answering a few of our questions about her encounter back on March 21, 1973. The story of her encounter with that unknown craft was pretty much the same as documented by multiple investigators over the past forty plus years.

Eddie Doyle Webb
October 3, 1973, 6:30am

Eddie Doyle Webb age 45, of Greenville, Missouri, was a truck driver in Cape Girardeau, Missouri for Sam Tanksley Trucking, Inc. His wife, 47-year-old Velma Mae was his trucking business partner. On the morning of October 3, 1973, the pair was driving a tractor-trailer rig south on interstate 55 in southern Missouri. Just before dawn at 6:30am, Webb was driving, and Velma Mae was asleep in the cab when something quite unexpected occurred. They were just south of the Jackson exit at Wedekind Park when the UFO incident occurred. Webb stated that a turnip-shaped object which appeared to have three sections, covered both lanes of Highway 55. The top and bottom looked to be made of aluminum and were spinning, the center section was stationary, and the craft was covered with glittering red and yellow lights.

Webb woke his wife, but she did not see anything out of the rear-view mirror. He decided to stick his head out of his window to try and see if there was anything behind them. Seconds later, he was blinded by a bright flash which he described as feeling like a ball of fire which had struck him in the head and face. The blindness was only temporary, so within a few minutes he was able to see with blurred vision but only three feet into the distance.

Webb's glasses, which he was wearing, were damaged by the heat. Under a microscope, the plastic frame appeared to have been heated internally. Dr. Harley Rutledge who inspected the glasses, stated that from the location of the damage, the destruction did not come from directly in from of him but from slightly to the left. This would verify Webb's statement that he was looking out his window toward the rear of the truck. The glasses were melted, slightly twisted, and displayed bear pit marks which resembled craters. The craters were caused by bubbles coming from the inside of the plastic to the surface and then bursting. Webb said that for several hours after he was hit by what he described as a hot flash; his sight was limited to eight to ten feet. His face was also covered with reddening blisters.

Webb's wife Velma Mae drove the rig the remainder of the way to their destination which was in Cape Girardeau. After they arrived, checked in and informed their boss of the incident, a company relief driver, drove Webb to the local hospital for treatment. The truck he was

Location where Eddie Webb had an encounter

Photos: Debbie Ziegelmeyer

Debbie Ziegelmeyer points to the location where Eddie Webb had his encounter

driving, a 1973 Peterbilt with a 40-foot Dorsey trailer refrigerator unit, did not seem to experience any damage. However, Webb's eyes were damaged permanently, being diagnosed as 15 degrees off.

The investigation into the incident was conducted by Sargent Edwin E. Wright of the State Highway Patrol; Edward O'Herin, New Madrid, Missouri, MUFON State Section Director for Southeast Missouri and Attorney at Law; Dr, Harley Rutledge; and MUFON Deputy Director-Administrator John Schuessler. Edward O'Herin, Attorney at Law, filed a Workman's Compensation Lawsuit on behalf of Eddie Webb who was driving a truck for Sam Tanksley Trucking, Inc. at the time of his injury and therefore was entitled to Workman's Compensation benefits. Because of the nature of the injury and circumstances leading up to the injury, Eddie Webb was awarded his Workman's Compensation benefits and to date may be the only person injured by a UFO incident who has.

from: The Southeast Missourian, Oct. 5, 1973, p. 1.

The weather
Cape Girardeau — Fair and cool tonight with lows upper 40 to low 50s. Saturday partly sunny with highs in upper 70s to low 80s.

THE SOUTHEAST MI

Vol. 70-No. 3 24 Pages CAPE GIRARDEAU, MO. FRIDAY, OCTOBER

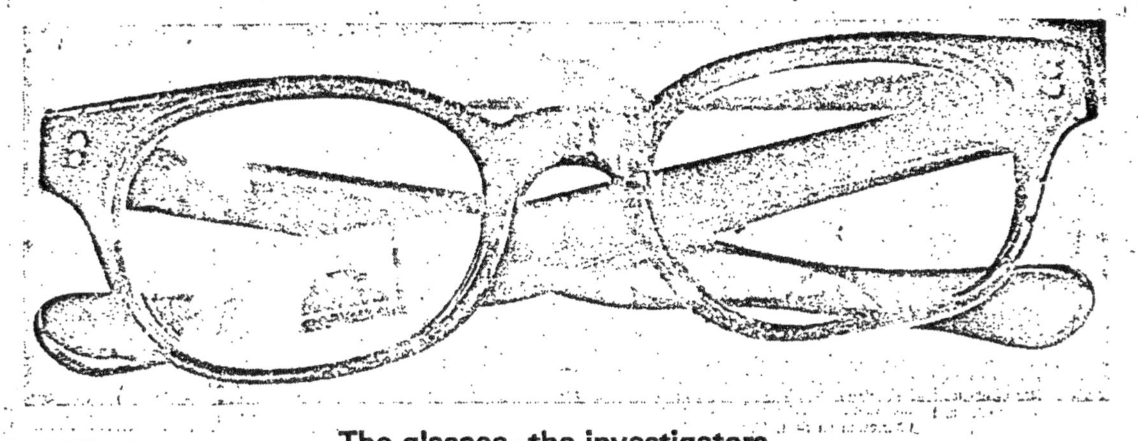

The glasses--the investigators

Investigation today centered around a pair of eyeglasses which were worn by Eddie Webb, 45, of Greenville, who claims he saw an unidentified flying object and was hit in the face by a ball of fire which blinded him Wednesday near Cape Girardeau. Above are the glasses shown slightly twisted and with pit marks probably caused by some form of heat. Below, principal investigators in the case, Sgt. Edwin E. Wright of the State Highway Patrol and Dr. Harley D. Rutledge, head of the SEMO State University physics' department, confer Thursday as Dr. Rutledge prepared to start a series of tests on the glasses.

Distorted glasses attention-getter

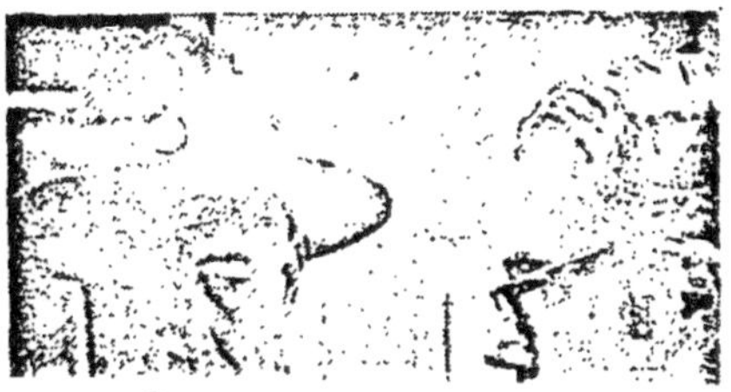

A
in

By LEE
Associated
BALTIMORE
Justice Depart
responds today
dent Spiro T. Ag
stop the federal
vestigation of hi
A lengthy gov
expected to reb
tention that it is
a vice president
nal offense.
Agnew's law
to halt the p
week the Cons
indicates that
must be imp
House and co
Senate before h
a criminal cou
requested a H
tion, but Speal

Co

The Army
Engineers is

would be illegal.

UFO blinds motorist?

CAPE GIRARDEAU, Mo. (UPI) — A physics professor investigating a reported attack on a truck driver by an unidentified flying object (UFO) said today the victim's eyeglasses were damaged by internal heat from an unknown source.

In the latest of a series of recent UFO sightings in southeastern Missouri, Eddie Doyle Webb, 45, of Greenville, Mo., was blinded for several hours after the incident. He is recovering his vision, but intends to visit an eye specialist at Barnes Hospital, St. Louis.

Webb said he was driving a tractor-trailer rig about dawn Wednesday when he saw a bright light or aluminum object in the air behind him, "coming up real fast."

He awakened his wife, Velma Mae Webb, 47, who was asleep in the cab, he said, but she didn't see anything.

"Then, I stuck my head out of the window and a large ball of fire struck me in the face," Webb said. "My glasses fell off and I couldn't see. But I got the truck stopped."

Mrs. Webb said her husband screamed, "Oh, my God! I'm burned! I can't see!"

One of the lenses of his glasses fell out of the plastic frame which was warped. Mrs. Webb, who serves as a relief driver at times, drove him to a hospital.

Sgt. Ed Wright of the Highway Patrol took Webb's glasses to Dr. Harley Rutledge, head of the Southwest Missouri State University physics department, for an analysis.

Rutledge, who has been working for six months to attempt to identify mysterious flying objects, said he put the glasses under a microscope and "it appeared they were heated internally.

"The plastic apparently got hot and the mold came to the surface. The heat warped the plastic, causing the lens to fall out."

Wash. State.

2A Everett Herald, Friday, October 5, 1973

Strange Flying Objects Blinds Truck Driver

Jefferson City (UPI)--Col. Sam S. Smith, superintendent of the Missouri Highway Patrol, said Thursday his men had received strange reports of actions of a flying object in the Missouri Bootheel this week.

The teletype message relayed to Smith at Patrol general headquarters from the Poplar Bluff troop said:

"About daylight on the morning of October 3, 1973, south of Jackson, Missouri on I-55 near the 34 and 51 exit, Eddie Doyle Webb, age 45, of Greenville, Missouri, was driving a tractor-trailer and his wife, Velma Mae Webb, was asleep in the cab. Eddie Doyle Webb saw a light on a large aluminum object in the air behind him. He awakened his wife, but she didn't see anything.

"Then, Eddie stuck his head out of the window and a large ball of fire struck him about the head and face. Subject's glasses and one lens broken out and the plastic frames were distorted from heat or acid of some type. Subject got the truck stopped in middle of road and his wife drove him to Southeast Hospital, Cape Girardeau, Missouri. Subject is blinded and is being examined by Doctor Cozine, eye specialist."

Col. Smith said the patrol has not turned up any unidentified flying objects or been able to explain the occurrence.

CLEVELAND PRESS - 10/5/73

UFO releases mystery heat

CAPE GIRARDEAU, Mo. — (UPI) — A physics professor investigating a reported attack on a truck driver by an unidentified flying object (UFO) said today the victim's eyeglasses were damaged by heat from an unknown source.

In the latest of a series of recent UFO sightings in Southeastern Missouri, Eddie D. Webb, 45, of Greenville, Mo., was blinded for several hours after the incident. He is recovering his vision, but intends to consult an eye specialist.

Webb said he was driving a tractor-trailer about dawn Wednesday when he saw a bright light or aluminum object in the air behind him, "coming up real fast."

He awakened his wife Velma, 47, who was in the cab, he said, but she didn't see anything.

"Then I stuck my head out of the window and a large ball of fire struck me in the face," Webb said. "My glasses fell off and I couldn't see. But I got the truck stopped."

Mrs. Webb said her husband screamed, "Oh, my God! I'm burned! I can't see!"

One lens of his glasses fell out of the plastic frame, which was warped. Mrs. Webb, who serves as a relief driver at times, drove him to a hospital.

Sgt. Ed Wright of the Missouri Highway Patrol took Webb's glasses to Dr. Harley Rutledge, head of Southwest Missouri State University's physics department.

Rutledge, who has been working for six months to attempt to identify mysterious flying objects, said he put the glasses under a microscope and "it appeared they were heated internally."

"The plastic apparently got hot and the mold came to the surface. The heat warped the plastic, causing the lens to fall out."

Illinois State Register, Springfield, Friday, Oct 5, 1973

UFO blinded me, truck driver says

CAPE GIRARDEAU, Mo (UPI) —A physics professor investigating a reported attack on a truck driver by an unidentified flying object (UFO) said today the victim's eyeglasses were damaged by internal heat from an unknown source

In the latest of a series of recent UFO sightings in southeastern Missouri, Eddie Doyle Webb, 45, of Greenville, Mo. was blinded for several hours after the incident He is recovering his vision, but intends to visit an eye specialist at Barnes Hospital, St Louis

Webb said he was driving a tractor-trailer rig about dawn Wednesday when he saw a bright light or aluminum object in the air behind him, "coming up real fast "

He awakened his wife, Velma Mae Webb 47, who was asleep in the cab, he said, but she didn't see anything

"Then, I stuck my head out of the window and a large ball of fire struck me in the face," Webb said "My glasses fell off and I couldn't see But I got the truck stopped "

Mrs Webb said her husband screamed, "Oh, my God! I'm burned! I can't see!"

One of the lenses of his glasses fell out of the plastic frame which was warped Mrs Webb, who serves as a relief driver at times, drove him to a hospital

Sgt Ed Wright of the Highway Patrol took Webb's glasses to Dr Harley Rutledge, head of the Southwest Missouri State University physics department, for an analysis

Rutledge, who has been working for six months to attempt to identify mysterious flyng objects, said he put the glasses under a microscope and "it appeared they were heated internally

"The plastic apparently got hot and the mold came to the surface The heat warped the plastic, causing the lens to fall out "

Rutledge said he planned more tests on the glasses He said there appears to be "some residue which we hope to put through some chemical tests "

In Tupelo Miss , police reported for the third consecutive night the sighting of multicolored UFOs Thursday

Wash. Star-News 10-5-73

GLASSES MELTED

UFO Blinded Trucker

CAPE GIRARDEAU, Mo. (UPI) — A physics professor investigating a reported attack on a truck driver by an unidentified flying object said today the victim's eyeglasses were damaged by internal heat from an unknown source.

In the latest of a series of recent UFO sightings in southeastern Missouri, Eddie Doyle Webb, 45, of Greenville, Mo., was blinded for several hours. He is recovering his vision.

Webb said he was driving a tractor-trailer rig about dawn Wednesday when he saw a bright light "coming up real fast."

He awakened his wife, Velma Mae Webb, 47, who was asleep in the cab, he said, but she didn't see anything.

"Then, I stuck my head out of the window and a large ball of fire struck me in the face," Webb said. "My glasses fell off and I couldn't see. But I got the truck stopped."

MRS. WEBB said her husband screamed, "Oh, my God! I'm burned! I can't see!"

One of the lenses of his glasses fell out of the plastic frame. Mrs. Webb, a relief driver at times, drove him to a hospital.

Sgt. Ed Wright of the Highway Patrol took Webb's glasses to Dr. Harley Rutledge, head of the Southwest Missouri State University Physics Department, for analysis.

Rutledge, who has been working for six months to attempt to identify mysterious flying objects, said he put the glasses under a microscope. "It appeared they were heated internally," he said. "The plastic apparently got hot and the mold came to the surface. The heat warped the plastic, causing the lens to fall out."

RUTLEDGE SAID he planned more tests on the glasses. There appears to be "some residue which we hope to put through some chemical tests," he said.

Meanwhile, at Tupelo, Miss., police reported for the third consecutive night the sighting of multicolored UFOs yesterday.

The Lee County sheriff's office said two deputies told of seeing brightly lighted objects in the Tupelo area and that similar reports had come from sheriff's departments in neighboring, Pontotoc and Itawamba counties.

"One of them was a rotating white and red light and another deputy saw a white light about the size of a table top with red and yellow lights behind it," a dispatcher at the Lee County sheriff's office said.

Chapter 3

The Initial Investigation Begins

By Debbie Ziegelmeyer

A team of investigating experts and consultants were brought together in Piedmont on March 31, 1973, by Missouri State Representative Jerry T. Howard for a weekend of witness interviews followed by a group review of information obtained to date. Making up this special investigating team were Dr. J. Allen Hynek, Chairman of Northwestern's Astronomy Department and a consultant for the U.S. Airforce on the UFO phenomenon for over twenty years; Ted Phillips, Sedalia, Missouri, MUFON State Section Director for the area and Secretary of the Mutual UFO Network; Edward O'Herin, New Madrid, Missouri, MUFON State Section Director for southeast Missouri; Robert Kewney, Quincy, Illinois, Chairman of Tri-State UFO Study Group; and Walt Andrus Jr., Director of the Mutual UFO Network. Clearwater High School basketball Coach and witness to the event Reggie Bone, arranged for a substantial number of witnesses to recent local UFO events including himself and basketball player Cary Barks, to gather at his home to be interview by the investigating team. The investigators were accompanied on almost every witness and on-site visit by locals including Police Chief Gene Bearden; Mayer Roy Anderson; Dennis Hovis, Manager of Radio Station KPWB; Harold Ellinghouse, reporter for the Piedmont newspaper; Dennis Keeney, announcer for KPWB radio; and Jerry Howard.

The incident reported by Reggie Bone and the members of his basketball team which were present, was classified as a UFO but officially categorized by the team as "nocturnal lights" within approximately two hundred yards.

Later that day the team would move to the home of Mrs. Maude Jefferis to review and analyze photographs she and local student Gary Sutton had taken.

Clearwater Lake
Photo: Debbie Ziegelmeyer

Gary Sutton was not available to answer inquiries concerning the circumstances behind what led him to photographing the objects, so the team was not able to come to any reasonable conclusions.

Maude Jefferies

The team of investigators studied an original print of a gyroscope shaped object which had been published in several newspapers and taken by Maude Jefferies. The two St. Louis newspapers had "cropped" the original picture removing the mercury vapor yard light form the left side of the shot. Mrs. Jefferies assisted the team in conducting an experiment to duplicate the exact conditions under which the original photograph was taken. By conducting this reenactment, the team could determine if the gyroscope shaped object was indeed a UFO. It was decided to take two separate photos making two separate exposures of ten and twenty minutes each. These were taken in the same direction as the original submitted photo which was a view from Mrs. Jefferis patio of her landscaping and pond under an overcast sky. Mrs. Jefferis developed these photos in her own darkroom and both negatives displayed the same gyroscope shaped disc in the upper right corner. The two photographs did not include star trails due to the cloudy overcast sky, and the team observed no UFOs during the two exposures. It was determined that the source of the light was "lens flare."

The Farmington Airport where Harley Rutledge's team did part of their investigation.

Photos taken by Debbie Ziegelmeyer in 2022

12A St. Louis Globe-Democrat Fri., Mar. 23, 1973

St. Louis Globe Democrat Impact

The folks in Piedmont, Mo., say strange things are happening in the sky above their Ozark town. Arrow at right indicates one of the multi-colored spinning objects one resident thinks must be a 'newfangled invention the goverment is playing tricks with.' Streaks in the time exposure photograph indicate stars.

'It came out of a lake'

And I looked, and, behold, a whirlwind came out of the north, a great cloud, and a fire infolding itself, and a brightness was about it, and out of the midst thereof as the color of amber, out of the midst of the fire.

Ezekiel: Chapter 1, Verse 4

By CHARLES J. OSWALD

Globe-Democrat Staff Writer

PIEDMONT, MO. — Clyde Morris thinks those strange objects seen regularly in the sky over this little Ozark town are not Biblical but some "newfangled invention" the federal government is playing tricks with.

Omer Fears thinks the weird, bobbing lights may be marsh gas or an experimental device in search of a lost silver mine but he won't commit himself until he actually catches a glimpse of it.

As he says, "I'm from Missouri and I'm going to have to see it before I believe it."

THE THEORIES were presented over a breakfast of country bacon and fried eggs at a downtown cafe here where the primary topic of conversation was the unidentified flying objects seen almost every night for the last month hovering over the Ozark Mountains.

Hundreds of residents of this leadbelt town 126 miles south of St. Louis claim to have seen the objects and some admit they are a little uneasy about the recuring lights in the sky.

Miss Jean Coleman and Miss Cathy Leach were driving across Clearwater dam about 9 p.m. Wednesday when they reported seeing the object surface from the reservoir side of the lake and take off.

The two women said the water took on a reddish glow and a disc-shaped object surfaced. It made no sound as it rose from the lake and soared over a mountain top.

The object turned from red to multi-colored spinning white, green, red and amber lights before it disappeared, they told authorities.

THE FIRST SIGHTING was made by Piedmont High School Coach Reg Bone and five athletes as they drove along the lonely road from Elsinore to Piedmont following a basketball game on the night of Feb. 21.

A bright light followed their car then stopped and hovered over a field about 100 yards from the road where the coach and boys were able to distinguish four intense amber, green, white and red lights.

Without a sound the object rose straight in the air and disappeared over a mountain at a high rate of speed, the witnesses said. There was no sound, no turbulence and no heat from the object.

Since then hundreds of residents have reported sighting the strange objects hovering over Clearwater Lake, shaking houses and skimming across the country side.

Gene Bearden, chief of police, said he's seen the lights and received hundreds of reports but he doesn't "know what to think of it."

"My first impression was that it is something McDonnell-Douglas or the government is experimenting with. Obviously they haven't harmed anyone and don't seem to pose a threat," Bearden said.

Ken Johnson and his family were driving up the switchback road leading to their home overlooking Clearwater dam at 9:30 p.m. March 12 when they saw a bright white light hovering about 800 feet above the lake.

Johnson, a licensed pilot and owner of a boat dock near the dam, said when they got to the top of the hill the object seemed to change its course and come at them.

"IT WAS ABOUT an eighth of a mile from the dam when it disappeared in mid-air like somebody turned off a light bulb," he said.

Dozens of witnesses interviewed separately gave similar descriptions of the phenomenon which they described as a bright, white or amber light that suddenly and mysteriously blinks off.

Wesley Howard and his wife, Gladys, their daughter, Jennie, and Mrs. Opal Faulkner were driving along highway 34 near the Black River about 9:30 p.m. March 15 when they saw a weird amber colored object "beaming rays" over the tree tops.

"It appeared to be about the size of a bushel basket from where we were. It glowed for about 20 seconds and then blinked out," Howard said.

Tom Martin, 14, and his friend, Kevin Kinard, 16, were camping on top of Bluff View overlooking the lake with eight other Explorer Scouts on March 19 when the boys saw a circular shaped glowing object skim across the water and disappear.

LATER THE SAME WEEK, five men fishing in the pre-dawn darkness on the lake started their outboard engines and headed for shore after seeing several disc-shaped objects hovering over the water.

Strange bubbles have been reported coming from the lake where Hayden Hewes, director of the International Unidentified Flying Object Bureau, Oklahoma City, believes the craft may hide during the daylight hours.

A farmer on the outskirts of town was watching television one night last week when he heard "a terrible rumbling sound and the whole house began to shake."

"At first I thought it was my dog scratching himself and shaking the floor," he told friends. "Then the television started acting up and went off."

"I went out to see what had happened when I looked up I saw a bright light hovering about 80 feet in the air over the house. Then it moved away and disappeared," he said.

MRS. MAUDE JEFFERIS, who teaches photography at the high school, claims to have taken time exposure photographs of several flying objects from the porch of her home behind the Pine Hill Drive-In Theater.

The pictures, taken by a Crown Graphic camera with a 135 m.m. lens at F 4.7 with a 10 minute exposure on Royal Pan 4 by 5 inch film from 11 to 11:10 p.m. on March 14, shows an object shaped like a child's top or gyroscope hanging in the western sky.

"When you look at it through binoculars it appears to rotate like a gyroscope," she said. "Closer observation revealed spinning or swirling white, amber, green and red lights."

Mrs. Jefferis said she has seen as many as four objects in the sky at one time. "They seem to blink signals to each other then speed off in different directions and disappear."

WHILE THE SIGHT of strange objects in the sky might raise goose pimples on the necks of some timid souls, the pragmatic Ozark people are taking things in stride.

"You'd think if the federal government didn't know what these things were they would send the Army in to investigate them. There ain't been one federal agent in town," Fears said.

"You reckon the American government has got something we don't know about?" Morris said, mopping up the gravy from his plate.

Mrs. Maude Jefferis said she took the photo on the top of this page. 'They (the obects) seem to blink signals to each other, then speed off in different directions and disappear,' she said.

Ken Johnson, left: 'It disappeared in mid-air like somebody turned off a light bulb.' Police Chief Gene Bearden, center: 'My first impression was that is something McDonnell-Douglas or the government is experimenting with.' Omer Fears, right: 'I'm from Missouri, and I'm going to have to see it to believe it.'
—Globe-Democrat Photos by Jack Fahland

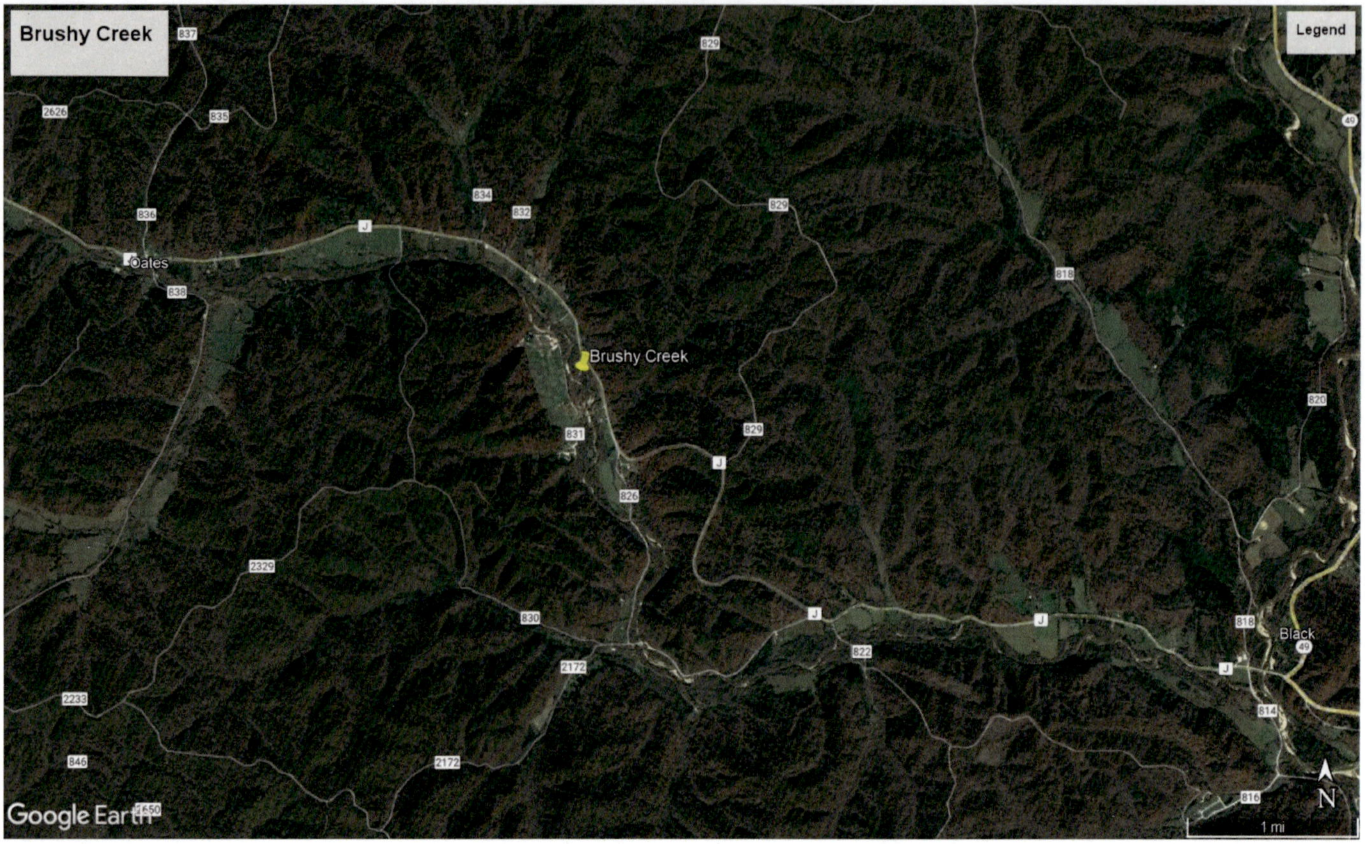

Brushy Creek area Source: Google Earth

Citizen Band Radio Alert

An alert went out via citizen band radio of a bright unidentified light. This light was reported to be seen by multiple witnesses on a farmer's property just beyond a pond. At the time of the sighting, the sky was overcast with a misty precipitation. These conditions all but eliminated the possibility of the object being a star. The main two witnesses involved, two gentlemen, refused to have their testimony recorded and their names to be used by the press.

On-Site

The next step in the investigating process was to do an on-site visit of several of the witness vantage points before the sun set. This would enable the team members to have a better perspective of area hills, valleys, lake, streams, pond, and roads. Within just a few days' time span, local law enforcement had received over five hundred calls from area citizens of unidentified lights or aircraft. While on location, some of the witnesses came forward and were interviewed by the team members.

February 25, 1973

Dennis Hovis interviewed two additional young men, nineteen-year-old Pat, resident of Piedmont, and his friend Will, who resided in Vulcan, a small town fifteen miles north of Piedmont. The two described a sighting they had on the night of February 25, 1973. The young men were on their way to Arcadia driving down the straight stretch of road just before Tip Top when they observed a strange red light. The light, which was just above the treetops, had three to four prongs on it and was jumping up and down, in and around, and behind the group of trees. The red light reappeared from behind the trees and was now about the size of four cars put together with two additional smaller red lights. The young men now believed they were looking at a strange craft with a dark square structure, prongs coming out from the bottom of it, and was approximately four hundred yards away. There was no sound, but they both stated that they were driving as they watched it which was for about ten minutes. Their vehicle did not experience any EMF effects, and they didn't notice any unusual smell in the area. The craft disappeared once again behind the trees. When they finally got to Tip Top for a better view, they stopped their car hoping to see it, but it didn't reappear.

March 12, 1973

A schoolteacher and her daughter were in the Brushy Creek area when they observed three strange blinking objects in the sky. The light moving north to south blinked out, the second light was moving east to west, and the third and closest object was a red ball of light 30 degrees above the horizon.

March 12, 1973, Monday 9:30pm

Ken, aged thirty-five who lives in the park which overlooks Clearwater Lake has a boat dock on his property. He was in the air force for four years, worked with radar, and is a private pilot which makes him a very credible witness when describing something out of the ordinary he saw in the sky. Ken and his wife saw a bright white light while driving towards Clearwater Lake dam on Highway HH. The sky was clear, there was no visible moon and no wind, it was a very calm evening. Ken and his wife were a quarter mile east of the dam when they noticed a white light moving north to south at approximately 3,000 feet altitude. The light moved towards the lake but when it was centered over the main part of the lake, turned towards their vehicle, and seemed to get brighter and larger. Still at approximately three miles from the couple, it made a 90-degree right turn then a left turn moving directly towards them. Ken stopped the car and attempted to take a picture but before he could focus the camera, the light "went out." At this point the light was 300-800 feet in altitude just one mile away. Ken was not able to give an estimate of size, but the color and density resembled a heat lamp or sun lamp. The light had a glowing effect, no beam, was white and steady never flashing, no sound, no EMF effects to their vehicle, and traveled approximately one hundred miles per hour. The duration of the couple's sighting was approximately two minutes.

Ken also described speaking to a camper who had seen what she thought was the same light, but earlier than his encounter and moving underwater. He also spoke with additional campers there at the Clearwater Lake Park, who were from St. Louis. They described a strange light they had seen from their campsite. Their campsite was on the ridge above the boat dock so they could see that the light was not a boat. The light was right underneath the water and from the reflection in the water seemed to be approximately two feet long. It was just out from the edge of the roadway moving from the cove out towards the middle of the lake. There were two ladies and seven children in the group of campers, two of the children were teenagers. One of the teens grabbed a flashlight to shine at the underwater object. When he did, the light went out and never reappeared.

March 21, 1973, 11:30pm

Carl, a Real Estate Broker aged forty-six and friend Ronnie, witnessed a white light, which was strobing but not flashing. Carl through binoculars, could see red, green, and white flashing lights. The craft was

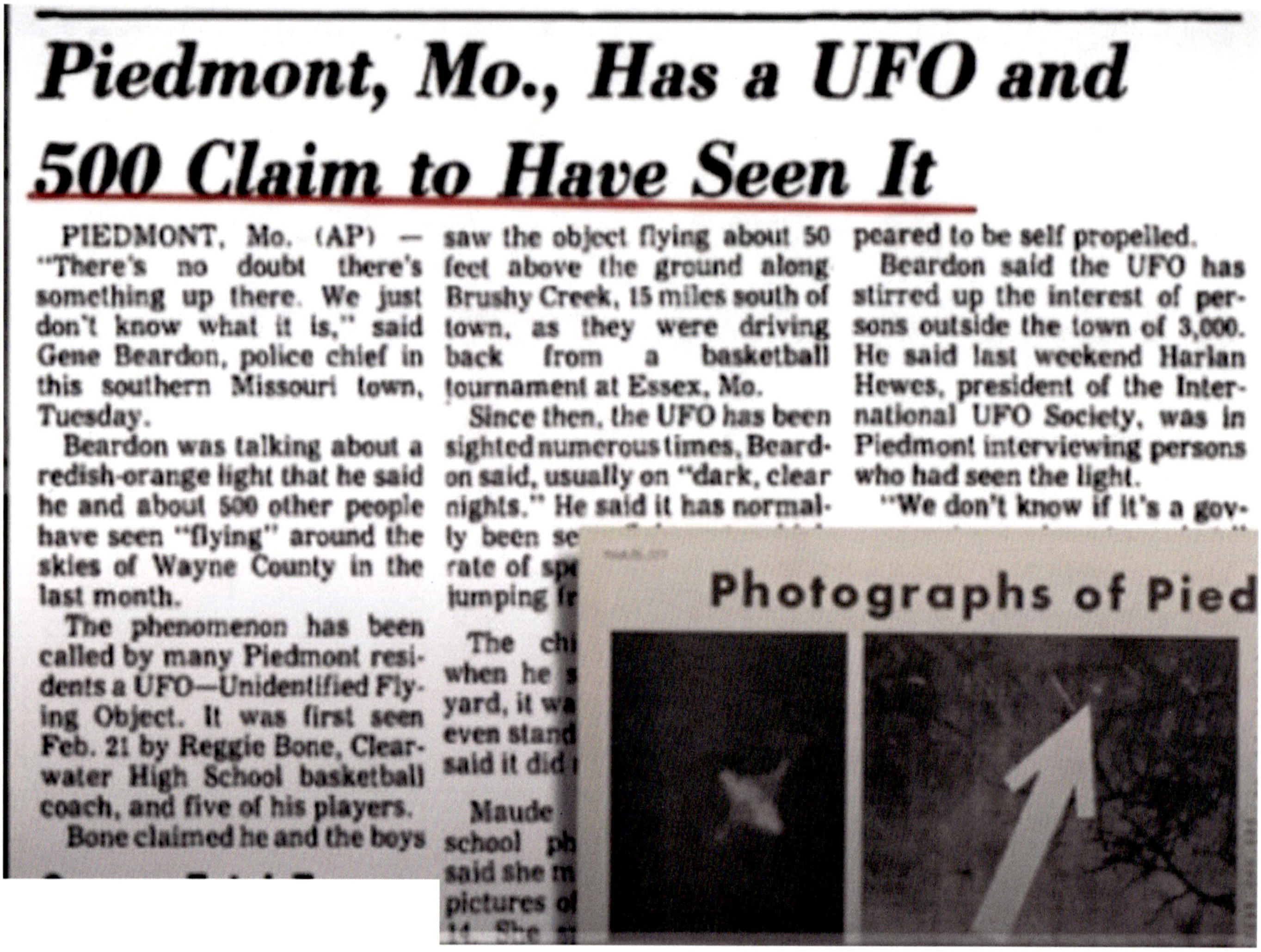

Piedmont, Mo., Has a UFO and 500 Claim to Have Seen It

PIEDMONT, Mo. (AP) — "There's no doubt there's something up there. We just don't know what it is," said Gene Beardon, police chief in this southern Missouri town, Tuesday.

Beardon was talking about a redish-orange light that he said he and about 500 other people have seen "flying" around the skies of Wayne County in the last month.

The phenomenon has been called by many Piedmont residents a UFO—Unidentified Flying Object. It was first seen Feb. 21 by Reggie Bone, Clearwater High School basketball coach, and five of his players.

Bone claimed he and the boys saw the object flying about 50 feet above the ground along Brushy Creek, 15 miles south of town, as they were driving back from a basketball tournament at Essex, Mo.

Since then, the UFO has been sighted numerous times, Beardon said, usually on "dark, clear nights." He said it has normally been se… rate of sp… jumping fr…

The chi… when he s… yard, it w… even stand… said it did…

Maude… school ph… said she m… pictures o… 14. She…

…peared to be self propelled.

Beardon said the UFO has stirred up the interest of persons outside the town of 3,000. He said last weekend Harlan Hewes, president of the International UFO Society, was in Piedmont interviewing persons who had seen the light.

"We don't know if it's a gov-…

Clipping from area newspaper. Credit: Wayne County Journal

completely silent, moving east to west, accelerated, reversed direction, and climbed in altitude. While over Spring Hill, the craft intercepted two red flashing lights coming from the south. The original light moved back towards the east.

Carl and Ronnie were traveling up highway 49 to Mill Springs. Just below Mill Springs, Ronnie spotted the light coming up from east to west. They got out of the car, both had binoculars and watched a flashing red light, a flashing green light, and a strobe light, like on an airplane. It moved across the sky and passed them between two and three thousand feet altitude. When it passed, the "real bright" flashing light went off. Carl had it in his binoculars and watched it go back the path that it came from, go straight up and bank to the right. After it leveled out again, the flashing light came back on and as it came back to the east, Ronnie spotted another light coming from south to north which was much lower, at about one thousand feet above the treetops. This craft had two flashing red lights and what amazed him was the speed and lack of noise. Carl stated that if it were an airplane "it would have to break the sound barrier." The two traveled to Brushy Creek, parked, and saw it back in the same area. Ronnie drove seventy miles an hour trying to get there to "head it off." It was heading in the direction of Clearwater Lake. The lights on the first craft were red lights, and a green light. Carl couldn't tell if it was two red lights and two green lights, but it had a "what I call in airplane terms a strobe light. This is a light that prevents mid-air collisions after night on air-planes." He stated that it could not have been a helicopter. "No noise whatsoever, if anybody doesn't think it is quiet down at Mill Springs about midnight at night well all we could hear was of course frogs and the regular other creepy things at night that holler around the swamps, well that is about all we could hear. The craft turned to go back from west to east, it made a turn, and I would say that it was as long as maybe as a small car and from looking at the red light and the green light, after it banked it just reminded you that was two arms." Carl is a pilot and could not explain what he and Ronnie had seen. It did not move like a normal aircraft, and it traveled at a much higher rate of speed than he knew possible. The sky was clear, but they could not see structure. The total duration of the second sighting was five minutes.

March 22, 1973, 4:30pm

Daylight Sighting

Two young men, Ron, from Piedmont, and Joe from Brushy Creek, were the only witnesses to a daylight sighting. The investigators went to Ron's house Sunday morning, but he was not at home. They were never able to personally interview either of the young men, but eventually heard their recorded interview which was made by Dennis Hovis.

On the afternoon of March 22nd at approximately 4:30pm, Ron and Joe, two southeast Missouri State University students were driving along Highway 34 eight miles outside of Piedmont when they both saw an oval-shaped craft in the sky. The craft, which was just above the trees, was "metallic" material with a flat underside and a dome on top.

March 23, 1973, 7:10pm CT

Leonard aged forty-nine, his wife, and their 13-year-old daughter Alma reported the following incident. While driving to Bluff View, they watched an airplane circle town crossing over the I.G.A. market. While looking up, the daughter noticed a bright blinking light. Leonard stopped the car and the three of them watched a bright light flashing ten to fifteen feet above the railroad tracks. Each time the light would flash it would move ten to fifteen feet higher. They watched this process for seven to eight minutes.

Additionally, Gary and Mark saw the airplane land and observed a second light at 7:45pm which continuously blinked as it slowly rose upward higher and higher in the sky. There was no sound being emitted from the light, which when reaching an altitude of five hundred feet, began responding exactly like an aircraft. The light was described by the men as moving straight up like a "STOL" vertical takeoff.

March 24, 1973

Small red balls of light were observed during an electrical storm but in a cloudless section of the sky. "Little puffs" were being emitted from these red balls of light. One of the balls of light descended to the ground behind a hill.

March 24, 1973

Five residents of Piedmont, Pete, his wife Betty, their friend Sherm and two girls, Dawn and Mary reported seeing a red flash moving very quickly across the sky. It descended behind the graveyard, took altitude again and turned white. It moved quickly behind the couple's house rotating as it moved displaying a green light, a red light, and a white light as it came around the nearby hill. It was down behind the trees when it turned white again flying extremely fast and low making no sound. The craft was large, and when it ascended from the trees again it was blinking in repetition in white, white, white, then a little red, then green and white but not like an airplane.

March 25, 1973, noon

Marjorie was driving across Clearwater Lake dam when she observed a large underwater streak coming across the water towards the dam. At the end of the wash something rose that resembled a man's hat. The hat-like object was darker than brown muddy water, extended twelve to eighteen inches above the surface, then descended back underwater. The incident lasted one minute.

March 26, 1973

The witness, Harley, a couple of days prior at 1:10am, watched through his window, a yellow ball of light flashing on and off moving from his barn to down the road. When asked if it could have been something in the water, Harley answered "no" it was close to the top of the trees moving extremely fast. He knew that it was not a helicopter because it was moving too low above the trees and there was no sound. There was just one yellow light seen, no others.

March 29, 1973 (dark) 7:10pm

Witnesses Glenda, husband and wife Bill and Kathryn, Dennis and Maude, while together on a local farm, reported seeing a large flashing light. Through binoculars the group could see green, red, and white lights. The light which changed colors and shift slightly, was photographed

by Maude. A second small red light seemed to come from the large light which moved rapidly to the south.

March 20, 21, 22, 1973

Three divers from East Side Divers Supply Company of Granite City, Illinois, Michael Henderson, Michael Sparks, and Ron Kilert, made three separate attempts three days in a row, March 20, 21, 22, to navigate Clearwater Lake for evidence of the underwater UFO. Unfortunately, an unusually heavy spring rainfall of seven inches that March had raised the water level several feet above normal. The lake was extremely murky, so the divers were unable to find or see anything at depth.

April 12, 1973

The Register, July 4, 1973, Santa Ana, California

A commercial pilot Kenneth Pingel, credited with more than 5,000 hours flying time, pursued a glowing disc-shaped UFO in southeastern Missouri. Pingel, accompanied by another pilot, Marvin Colyer, was approaching the airport at Farmington, Missouri, when he first sighted the white circular object off the left-wing of his single engine Piper Cherokee. ""The object gradually turned yellow and took on an orange "tingle," Pingel said. It did not appear to spin." The pilot stated that the UFO paced the plane": speed 150 mph, altitude: 3,500 feet. Heat waves, "like the kind you see on highways during a hot day, seemed to emanate from the object. As the fliers were making their final approach to the north runway, the object gave off a beam of white light. Then, it entered the plane's approach pattern." It looked like it was moving at us at a high rate of speed," the pilot said, "so I pulled back and flew toward it. It immediately stopped, reversed its direction, and flew away from us at a high rate of speed". Colyer and Pingel state that they chased the object for several miles at 140 mph. "We are both qualified for night operations, and I have flown many nights under all kind of conditions," Pingel said. "This was definitely not a star or any kind of aircraft."

April 20, 1973
Arkansas-Missouri Power Company
Subject: Switching Breaker Operations Record
Many locals reported power outages during UFO sightings. The Investigating Team requested information on line switching and breaker operations through the period of February 1, 1973, and April 8th. In response the following information was released:

"We have only three switch operations on our 34 KV line and all three were during daylight hours and on our breaker No. 890.
One on March 1st., 8:25 AM, March 6th., 8:39 AM and March 13, 3:20 PM.
According to our record of breaker operations during this period, we have had five on 890 and one on 891. Since the breaker operations are not checked every day, the following is the only information we can furnish:

Breaker 890
Between February 26 and March 5: (1)
Between March 13 and March 19: (1)
Between March 19 and March 26: (1)
Between March 26 and March 27: (1)
Between March 27 and' April 12: (1)
Breaker 891
Between March 26 and March 27: (1)
All charts are mailed to the Ironton office and from there are forwarded to the General office.

According to the power company records, no outages occurred due to UFO activity. However, the report did state the following: "Since the breaker operations are not checked every day, the following is the only information we can furnish." Which means that power outages could have occurred but were not detected or documented.

WJBC Radio Report: Monday April 22 or 23, 1973
On April 22 or 23, a private pilot on approach to the Farmington, Missouri airfield, spotted an orange disk which was in his path and coming near at a high velocity of speed. The pilot swerved and circled around behind the object which has slowed down. The pilot then proceeded to chase the object up to speeds of 140 miles per hour for a period of 10 – 15 minutes. There were no details listed in the report of what happened to the object after the short chase.

The Outcome
The special investigating team of Dr. J. Allen Hynek, Ted Phillips, Edward O'Herin, Robert Kewney, and Walt Andrus Jr., were one of the largest teams of UFO experts ever assembled to investigate a single area and series of events. Local MUFON investigator Edward O'Herin and KPWB radio station manager Dennis Hovis, relentlessly continued investigations into additional sightings for several months after the initial event. Locals gathered in open fields and parks every weekend for two additional months hoping to see something "other worldly." Many additional sighting reports came in, some which were thought by the investigating team to be airplanes or helicopters, but other reports of lights and craft dancing around in the night sky or traveling at high rates of speed were still deemed "unknown." Over the years local witness reports have significantly lessoned not due to the lack of

activity but because the area residents have grown accustomed to seeing strange unidentified lights, sometimes in groups, dancing in the dark night sky overhead and accept it as "life in Brushy Hill".

Just one month prior to the official beginning of the Brushy Hills investigation and during the height of the five hundred plus reported sightings, J. Allen Hynek made a public appearance in St. Louis, Missouri.

February 23, 1973
The UFO Experience; An Illustrated Lecture by J Allen Hynek
Marriot Motor Hotel, St. Louis Missouri
Sponsored by Northwestern University Alumni Club of St. Louis

Dr. J. Allen Hynek responses to questions posed by local media:

When asked about an eventual solution to the UFO enigma:

"A hundred years ago, if you had been the most brilliant physicist in the world, and Congress had given you a billion dollars to solve the problem of the Northern Lights, you couldn't have done it, because it was there, people could see it, because, simply, we didn't know enough physics yet. And we may be in a similar situation with the UFO phenomenon; it may be opening a totally new domain of nature. Just because its 1973 doesn't mean that we know the things that we're going to know about the universe in the year 4000."

Response to a question concerning the search for intelligent life in space:

"There seems to be bits of evidence, certainly undocumented and unproved, but there seems to be some evidence that, maybe life has come here. Now, NASA has in its charter that its mission is to go out and find evidence of life elsewhere; that's one of its very definite aims. It seems rather paradoxical to me that they won't give any effort at all to examining whether life elsewhere has already been here. The UFOs might possibly be that. I don't say they are, but suppose there's only one chance in a hundred, it would still be worth investigating it.

Statements by Hynek:

"There remains no doubt in my mind that a real UFO phenomenon, of some sort, exists, which may or may not have extraterrestrial origin. Indeed, that is the problem. Over the course of a score of years, both officially and unofficially, I have interrogated many hundreds of persons who claim to have witnessed the UFO phenomenon, and their aggregate evidence is so overwhelming, I have but two choices in evaluating their accounts: either to hold that these persons, adjudged responsible by all ordinary standards, are practiced prevaricators, or are suffering from some very serious mental disorder; or, to hold that they furnished us data that may be of decided, potential, scientific value.

Dr. J. Allen Hynek with Jacques Vallee

By United States Government - United States Government, Public Domain, https://commons.wikimedia.org/w/index.php?curid=1649744

Right: John Schuessler

Ted Phillips at the far left, with Adam Johnson (videographer) and the late Stanton Friedman during an investigation at the Marley Woods in Missouri (another paranormal hot spot).

Dr. Harley Rutledge examining Eddie Webb's damaged glasses.

Source: Weebly

Chapter 4

Harley D. Rutledge, Ph.D. Investigates

By Debbie Ziegelmeyer

Harley D. Rutledge was in his tenth year tenth year at Southeast Missouri State University at Cape Girardeau, serving the last nine as chairman of the Physics Department. His Ph.D. degree was in solid-state physics, photoelectric emission from the chemical compound strontium oxide. Although a bit of a skeptic on the UFO subject, his interest went back to books he had read on the subject in the 1950s. Curiosity got the better of Rutledge when he learned of the massive amount of UFO sightings in the Brushy Creek area, and Piedmont was only seventy-five miles away.

Harley Rutledge made the logical decision, to form a research group an investigation he named "Project Identification." Project Identification consisted of "a team of professionally qualified, highly trained researchers headed by Harley Rutledge, Ph.D. whose main objective is to investigate UFOs while the phenomena are in progress. The use of highly scientific instrumentation, a Questar telescope, whose magnifying can be changed without replacing the eyepiece, high quality camera equipment, electromagnetic frequency analyzer, and high frequency, low-intensity sound detectors, enables the team to calculate the UFO's actual velocity, distance, and size, as well as provide remarkable photographs of the real thing.

Project Identification officially began with a letter to University President Mark Scully from Harley Rutledge on March 27, 1973:

"After a bit of soul searching and discussion with colleagues, some of us in the department of physics have decided to investigate the strange lights appearing in the sky around Piedmont, Missouri. We plan to go as a team of scientists from the University, limited to me, Dr. Sidney Hodges,

Professor Milton Ueleke, and two students. The students are John Wilson and King Turnbull. Tentatively, we plan to take scientific recording instruments to the locale with the purpose of obtaining data if we are fortunate. The instruments include several astronomical telescopes rigged with cameras or light-sensing devices, a magnetometer, sound recorders, radios, and binoculars. As a matter of courtesy, I am informing you of our plans, as there may be publicity involving us as university representatives."

The actual involvement by the team began April 6, 1973, six weeks after Reggie Bone and his team's sighting when Rutledge, Ueleke and their wives drove into Piedmont at 5:30pm that Friday afternoon. The first person to greet them was Dennis Hovis KPWB radio, who began to inform them of some of the more important sighting incidents. That night the scientists staged in an area called Pyle's Mountain where they found a concrete slab which was perfect for mounting their Questar telescope. Less than an hour had passed when they saw the first of what would be five appearances of a small stationary light each time blinking out and reappearing in a different area in the sky. And so, it began, unexplained blue lights moving through the woods, white lights dancing in the night sky overhead, and unidentified fast moving white or amber lights traversing through the dark nighttime sky.

Monday, May 11, 1973, Rutledge put forth his plan, to establish a radio network of three teams of two persons each. One team had the task of watching for UFOs from their staging area of Pyle's Mountain; the second team was staged at the fire tower on Mudlick Mountain; and the third team would stage at the airport by the Cessna 150 ready to intercept if needed. If an unknown light or craft were seen by any of the locals, they were instructed to contact radio station KPWB in Piedmont who would radio the location to the team at the fire tower who would radio the team at Pyle's Mountain and the team at the airport.

One of the research teams more memorable moments in the Brushy Creek area occurred on May 11, 1973, while in flight to Piedmont from Cape Girardeau in a Cessna 182 at approximately 7:00pm. Jim Trickey was the pilot, Rutledge in the co-pilot seat with John Wilson and Drake Kambitch in back. Rutledge decided to have Trickey circle the area flying north, to set boundary lines for their area of search and cross Clearwater Lake on the return to Piedmont.

At 8:10pm, Rutledge spotted a small, pulsating light that had just appeared in the sky at 45 degrees many miles away. He watched the light for about ten seconds through binoculars when suddenly, a second light appeared which just "blinked on." Rutledge continued watching the two lights taking turns pulsating three or four times, then the other would seem to answer. Suddenly, a row of seven lights appeared in his field of view making this a group of nine white lights total. The seven lights took a position between the two pulsating lights. At that point, the initial two lights stopped pulsating.

The seven lights were not identical, three were closer and surrounded by a "peculiar, dark, cloud-like shimmering haze." The other four lights were farther away and seemed to "twinkle and Jiggle" in position. Rutledge pointed in the direction of the lights, Trickey changed course by 20 degrees right of north and increased his airspeed to 140 mph. Because they were in the back seat and out of field of view range, Wilson and Kambitch were

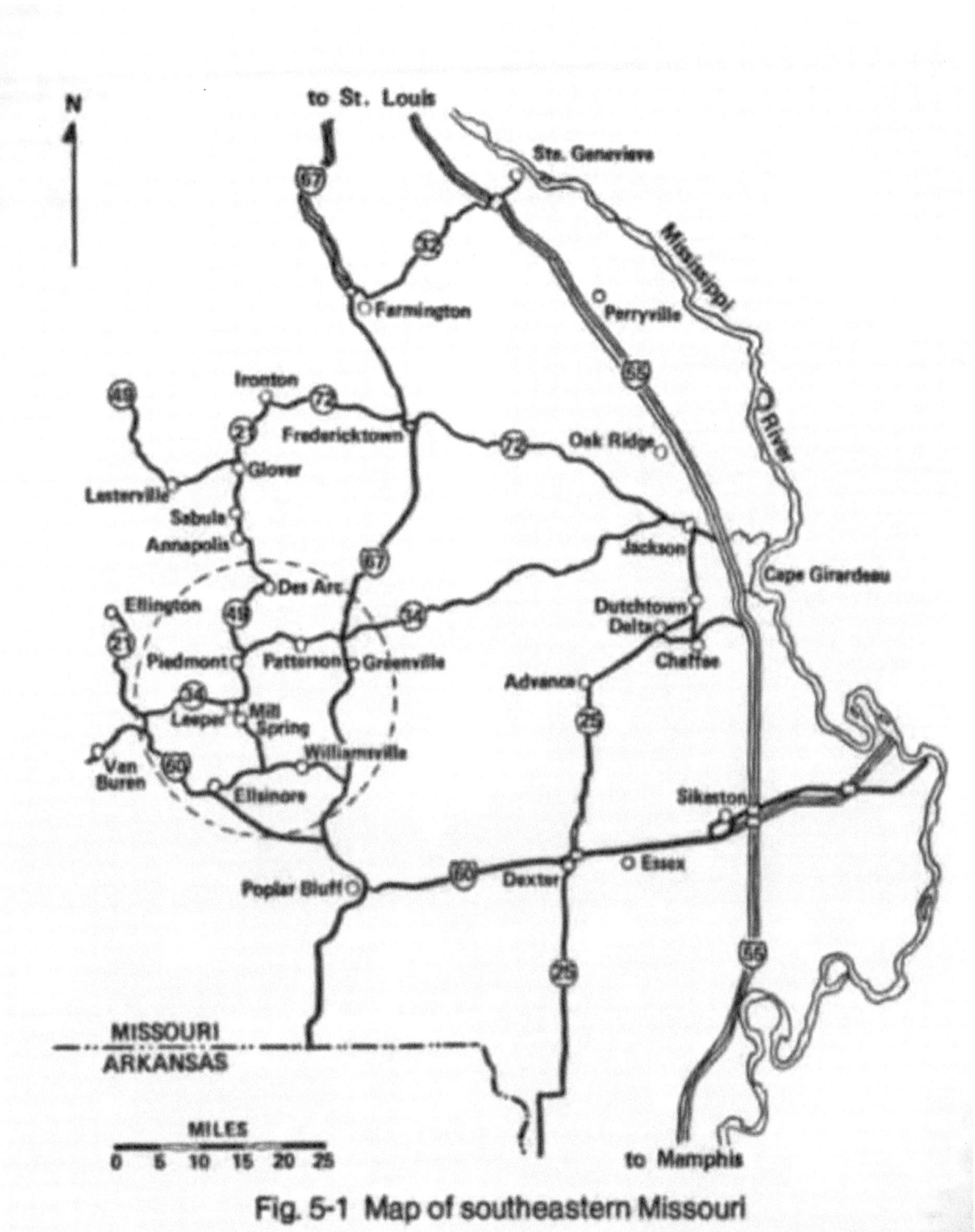

Fig. 5-1 Map of southeastern Missouri

The Investigation area as shown in the book “Project Identification” by Harley Rutledge, PhD

Project Identification may be viewed at www.archive.org at no charge

unable to see the lights which had just increased to ten total.

At 9:10pm, the team was a few miles east of Piedmont, flying east at 3,500 feet ninety mph. Wilson was keeping watch through the back window when out of nowhere a light was moving towards them at a high rate of speed. Wilson screamed "It's coming right at us!" When Trickey turned their plane to face the light head on, they lost sight of it. Wilson described the light he had seen as a "fluorescent-colored white light that flashed at least four times per second in equal time intervals. The motion of the light was jerky passing over their aircraft at less than one half-mile above." After their plane landed, Rutledge instructed the others to not discuss what they had seen with anyone but their immediately family members.

Although Piedmont was the primary headquarters, Project members had also set up equipment at Farmington airport, Williamsville, and in Sam Baker State Park. During the first week, on May 24′ 1973, 4:30pm, John Wilson and Harley Rutledge left the main group in Piedmont and set up equipment in Farmington in the airport for two consecutive nights of sky watching. They were led through the gate of the airport to the east side of the runway by Chuck Pingel, the owner's son, where they parked near the windsock as far as possible from the rotating beacon light. Rutledge placed a transistor radio on the car roof to listen to radio broadcasts and to serve a dual purpose, the possibility of experiencing electromagnetic interference. At 9:21pm shortly after an airplane had flown south of the airport, a stationary amber light appeared southeast of the airport. The light's brightness did not vary and by rotation of the camera lens, Rutledge was able to determine that the light was partially polarized. Soon after, an aircraft flew directly over the runway, headed in the direction of Farmington less than a mile away. Because of the low altitude of the aircraft, it's individual lights were easily seen. Pingel remarked that the aircraft sounded like a turboprop but moments later yelled "Look at that, he's got a red light on the right wing!" FAA regulations state that the light should have been green with the red light on the left wing. Wilson got an exposure but unfortunately the film was black and white.

Similar Recent Sighting by St. Louis Area MUFON Team:

During a Missouri MUFON investigation in the Saint Clair area on January 5, 2020, the team of myself, (Debbie Ziegelmeyer), Mel Van Vickle, and Russ Payne, witnessed this same phenomenon. We watched a ball of light come through the woods, take altitude, and suddenly display aircraft running lights. Mel yelled "Look, the red light is on the wrong wing! The aircraft would have to be flying backwards! Fortunately, our team does have this on video.

A few minutes later, what appeared to be the original hovering light reappeared in the same location. It had blinked off fifteen minutes prior. The light began to move and within three minutes had disappeared over the ridge to the northeast. As Rutledge looked northwest, he noticed a configuration of four lights which had just passed over head. From left to right the lights were white, red, red, and white and seemed to be attached to what resembled a huge wing. The craft was calculated to be 2,500 feet in altitude and six hundred feet across, or at 1,500 feet in altitude be 368 feet across. Later that evening on the way to Fredericktown, another strange strobe like light was seen dancing in the sky.

The team had planned to remain in Piedmont for three weeks but after only two, relocated to Cape Girardeau because of increased UFO sighting activity in that area. Over the next few years and several Missouri locations the team witnessed, photographed, and documented unidentified low moving balls of light, some white, some amber, and some blue. High altitude unidentified balls of light and streaks of light, were also seen, photographed, documented and their movement studied and analyzed. Some of these lights were seen moving irradicably across the night sky sometimes at a high rate of speed. Rutledge and his team investigated their experiences by means of science, using mathematical equations, charts, camera time, shutter speed, film exposure, and numerous measurements obtained from the equipment they had brought with them from the University.

At least thirty-five scientists, engineers, students, and lay persons had assisted in the field research at one time or another. Rutledge insisted that "lay reports" "not be accepted as scientific data." The term "lay reports" is used to describe a report either written or verbally, from a person who does not have a background in one of the sciences. Since this study and these investigations were being conducted in a scientific manor and sanctioned by the University, it was expected that results would be submitted in the form of scientific papers. "Lay reports" were welcomed by the research team as a means of triangulating a sighting or determining an active area.

Project Identification, which was originally planned to last three weeks, ran from 1973 to 1980. A total of 158 viewing stations were set up, 427 hours of sky watch time was documented, 620 viewers were on watch, 378 Project observers were used, 1,919 hours were logged by observers and 1,161 by Project members. Within the Project there were 157 sightings, 173 UFOs observed of which thirty-four were Class A sightings with 45 UFOs documented. Obviously, some sightings involved multiple craft or light sightings. Project areas of observation were from three major geographical areas: Cape Girardeau, Piedmont, and Farmington with additional sightings coming from Perryville and Sikeston.

Although the Project members believed science

would provide answers, it only produced more questions. Sightings are still occurring in the area to date, but reports are down. The residents have been living with the phenomenon for so many years that the events have just become a part of normal life in Brushy Creek.

News Publications

"People are Saying" article, volume 32 #17
22 December 1977

Dr. Harley D. Rutledge, head of the physics department of Southeast Missouri State University in Cape Girardeau, Mo., after more than four years of firsthand research and observation, has reluctantly admitted that he cannot debunk the stories concerning UFOs. He began as a skeptic in 1973 when it was suggested he investigate a rash of sightings of UFOs in southeast Missouri. The information he gathered included more than seven hundred photographs and is a result of hundreds of nights spent in the air, in fields, and on hillsides, accompanied, always, by other observers and equipment which include telescopes, cameras, and a spectrograph. Rutledge is laying his reputation and career on the line by saying there really is something up there that cannot be explained by conventional logic. "Now I know they are up there. What they are or where they came from, I have no way of knowing. But they are there, he said in a recent interview.

Rutledge himself has made more than 140 sightings, both in broad daylight and at night. At least twenty-five those sightings he labels "incredible." Three of the most convincing sightings came within two weeks of each other in May of 1973. The first, from the air, involved, what he calls a 45-second "metamorphosis" involving ten balls of light that had no business being where they were in the air near the small town of Piedmont, Mo. The second and third sightings, which he believes to be conclusive because he was so close to the objects, were made near Farmington, Mo. about two weeks later. On the night of May 24, Rutledge and several other observers spotted an object with a triangular shaped light pattern passing over them. Rutledge tentatively labeled the object as an airplane but said that later evidence discounted that possibility.

The next night, he said, he and his colleagues observed a huge object that had nearly passed over them before they spotted it. "It had four lights, two red and two white. "It was close enough that I could see parts of it through my 80-power telescope. It had a metallic skin, and I could make out a ribbed pattern in the red lights," stated Rutledge. Because of the delicacy of his position, Rutledge said he was reluctant to talk about the phenomena, but feels he has an obligation to tell the public what he knows.

Top Scientist Reveals... I've Made 25 Sightings of Intelligently Controlled UFOs
By Bernie Ward

A university professor investigating UFOs reveals he has sighted mysterious objects in the skies over Missouri including about twenty-five which he says were "truly incredible and unquestionably UFOs." "They are intelligently controlled. I won't back down from anyone on that statement," Dr. Harley D. Rutledge declared flatly. "They are not flying at random." Since 1973, when he first launched his investigation, Dr. Rutledge, head of the physics at full department at Southeast Missouri State University and former president of the Missouri Academy of Science, has made more than 140 UFO sightings. "Some were going speed instantly, and I also saw right-angle turns," said Dr. Rutledge, who used photography and scientific instruments to document a report he is preparing for scientific journals.

The professor's chilling conversion from skeptic to true believer came the night of May 11, 1973. With a pilot and two physics students, he was scanning the night sky with binoculars from the cockpit of a single-engine plane when he saw a strangely pulsing white light. "Another light came on, then a row of seven lights, and the three on the right had a kind of haze around them. Little beams of light seemed to shoot out from the center lights. "I have to admit that I had chills." A few minutes later he said one of the students screamed when he spotted an eerie flashing light closing fast on the plane, then abruptly disappeared. At 9:30pm on May 24, 1973, Dr. Rutledge and other scientists watched in awe from the ground as a spooky amber glow appeared below clouds at 7,000 feet, went off when a plane approached, then came on again for about three minutes. The team took photographs of the light. Later that night the observers beheld another astonishing sight. Moving silently across the sky was another object "about 300 feet across" with four long narrow lights reflecting off what appeared to be aluminum shaped around the back of something.

The next night, Dr. Rutledge said, another team miles away were dazzled by a round light that moved in their direction, then turned away. Dr. James Sage, then a Southeast Missouri professor of electronics, was surprised to find his spectrum analyzer used to measure electromagnetic frequencies was inexplicably jammed shortly after the light disappeared. Triangular measurements placed the object's speed at between 309 and 325 m.p.h., according to Dr. Rutledge. In the second week of June 1973, Dr. Rutledge sighted a bullet-shaped object flying in broad daylight. "It was round in front he remembered, "and square in back, and it "changed colors."

On Sept. 5, 1976, Dr. Rutledge made a daylight sighting right from the window of his office that he will never forget. "In the fraction of a second that I saw it," he remembered, "I had a glimpse of a disc shape with a flat bottom." John Schuessler, a project manager for the space shuttle, is impressed by the results of Dr. Rutledge's research. "It's been very scientific," he told the ENQUIRER. "And therefore, I consider it very good data.

Review about "Project Identification" on Amazon by "Bob" on October 25, 2001

First-Hand Experience: I was growing up in the town of Piedmont during all of the activity described. Dr. Rutledge's experience was a vindication of sorts for the whole town. Here was an admitted sceptic and a man of science saying that there really were things in the sky around Piedmont that couldn't be explained by natural phenomena.

I'm not a mathematical type at all, but I found Dr. Rutledge's use of formulas and his explanations fairly easy to follow. The photos were very interesting, too.

everyday

ST LOUIS POST-DISPATCH
SUNDAY, DECEMBER 4, 1977
SECTION 1-18L

Post-Dispatch Photos
By Gene Pospeshil

UFO investigators set up equipment at dusk on a farm road near Chaffee, Mo.

Dedicated Watcher Of Sky Mysteries

By Jim Creighton

Of the Post-Dispatch Staff

CAPE GIRARDEAU, Mo.

He had been talking about equipment. About the Navy-surplus rangefinder sitting in the center of his office window. About the magnetometer, the radiation detector and the spectrum analyzer on the shelves beside the desk.

Harley Rutledge stopped abruptly and leaned back in his chair.

"You know," he said, "you really can't sit here and accept what I'm telling you. Your conscious mind might, but not your subconscious.

"My feeling is that I'm not trying to prove this to anyone. I'm just telling you what I've seen. But perhaps the fact that you're hearing this from someone in my position might make you think about it a little more."

Harley Rutledge demonstrates the working of a rangefinder.

Top portion of page. Continued on next page...

Credit: St. Louis Post–Dispatch

you're hearing this from someone in my position might make you think about it a little more."

Rutledge is chairman of the physics department at Southeast Missouri State University. He has been for 13 years. He is 51 and married, with five children. He is respected by his colleagues. He used to pitch softball and, more recently, play a lot of bridge.

He has also, he says, seen more than 140 Unidentified Flying Objects.

A pickup, a station wagon, two cars and finally another station wagon were parked in a line down the rutted farm road near Chaffee, south of Cape Girardeau. Along the edge of the road, Rutledge, his 15-year-old son, Mark, and three members of the Southeast Missouri Astronomy Club had set up their sky-watching gear.

There were three eight-inch reflecting telescopes, two with cameras attached. The rangefinder was about halfway down the line, near the radar dish attached to the bumper of one of the station wagons. A camera with an 800mm lens was mounted on a tripod. On another tripod were side-by-side cameras, one with ultraviolet film. An infrared scope was lying on the hood of a car, and down at the end of the row was a battery of aircraft landing lights.

The sun was just going down, but it was already very cold, with a chilling wind cutting across the flat farmland. Rutledge had on a heavy coat, and the microphone from his tape recorder was fastened to one lapel. If something showed up, he could use the tape recorder and still have both hands free to operate a telescope or camera.

This was mostly a dry run, a demonstration of the equipment available for UFO searches, but Mark Rutledge went through the standard procedure for beginning an observation session. He punched on another tape recorder and recorded the date, the site and the members of the observing team. Then he taped the official time from a radio tuned to WWV, the National Bureau of Standards station in Fort Collins, Colo.

A big orange moon had edged over the horizon. Although it was barely 5 o'clock, it was already dark.

"We've got something up there, Harley. Must be a satellite, it's moving so fast."

"Where is it?"

"It's heading on southeast. I think it's popped out already. It was just a satellite."

For the last four years, starting not long after he reluctantly began an investigation of a "flying saucer" scare at Piedmont, Mo., in 1973, Harley Rutledge has been a dedicated UFO watcher. He is not completely comfortable in that role, but he has approached it with an intensity that he considers appropriate in a scientist.

Rutledge and his son Mark inspect the antenna for a spectrum analyzer.

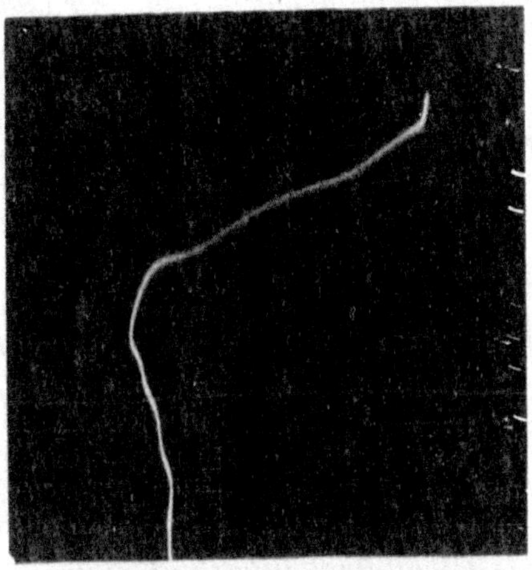

Mark Rutledge reports taking this photo from his yard last summer, using a tripod-mounted camera with an 800mm lens. It is purportedly a three-second exposure of the path of a bluish light that entered the camera's field on a diagonal path from the right. The slight dip at the beginning of the streak, Mark thinks, was caused by movement from the cable release. This photo is not one of those in his father's collection, the best of which, Rutledge says, he is saving for use with any papers he may publish in scientific journals.

What Rutledge has done, he thinks, is to demonstrate the "temporal spatial density" of UFOs. He says he has shown that there are sometimes enough of them around in one area at a particular time that a successful field study of the phenomenon can be undertaken.

It should be pointed out here that UFOs, if they exist at all, are not necessarily space ships piloted by LGMs (little green men) or BEMs (bug-eyed monsters). The extraterrestrial theory is popular, especially on what is perhaps unkindly described as the lunatic fringe, but UFO buffs say the things conceivably could be man-made or unexplained natural phenomena or even, as is increasingly suggested, some kind of psychic manifestation.

"People ask where they come from," said Rutledge. "Well, I don't know. I've got some pretty good ideas, but I don't share them with anyone."

With some unusual exceptions, most of the UFOs that Rutledge says he has seen fall into the category of nocturnal lights: hovering lights, lights that switch on and off, lights that jump about in startling fashion, groups of lights that burst on like flashbulbs, moving lights that make right-angle turns.

Many of these reported sightings came in a year of intensive field investigation that followed the Piedmont episode, a period in which a great many UFO reports were made in southeastern Missouri and southern Illinois. This was before Rutledge had all of his present detection equipment, but he says that the sightings were usually witnessed by other persons and were often photographed. He has more than 700 photos of odd lights, ostensibly UFOs.

Rutledge is right about one thing: No matter how convincing he himself may seem, no matter how obvious his sincerity, it is very difficult for most of us to accept the reality of UFOs. The idea is simply too far outside the limits of everyday experience, farther than we are conditioned to go.

He says that he was as skeptical as anyone else when he began hearing in late February 1973 about UFO reports from Piedmont, an Ozark community about 70 miles west of Cape Girardeau.

"Some of the things that I was reading I didn't like," he recalls. "I didn't like the way the Piedmonters were being treated by the media, and I was disturbed that some of the people there were truly frightened. Women were even carrying firearms at night."

By early April, Rutledge was wondering whether it might have a calming effect if someone from the university would drive over to Piedmont and look around. He wasn't sure that he wanted to be the one; he'd told a colleague about a week before that he "wouldn't touch that with a 10-foot pole," but in the end he decided he should.

On April 6, a Friday, Rutledge, astronomy professor Milton Ueleke and their wives drove over to Piedmont. They didn't expect to see anything, but they put a telescope and binoculars in the car. After dinner with some Piedmont friends of the Uelekes, they drove up to the city dump, a high spot where many of the sightings had been reported.

After about 45 minutes, an odd light appeared to the south, then blinked out. It reappeared three more times that evening, a well-defined image in the telescope, according to Rutledge. Car lights from some nearby road, Ueleke said. Rutledge didn't think so.

Rutledge's curiosity was aroused. The next Thursday, he and Drake Kambitch, a physics major at Southeast and a pilot, flew to Piedmont and began looking from the air for a road where the lights could have originated. The closest one was an old logging road, nearly 30 miles away.

Rutledge came back to Piedmont the next night and every Friday after that until the end of the school semester. He says he saw more unusual lights from other places in the area, including one streak of light that seemed to reach another light, then stop — something that people around Piedmont called an "intersection" of lights.

He still wasn't convinced, he says. But on May 11, another Friday evening, he underwent what he describes as a "45-second metamorphosis."

Rutledge, Kambitch, Jim Trickey and John Wilson flew to Piedmont that evening in a four-seater Cessna. Like Kambitch, Wilson was a physics major. Trickey, the pilot, worked for Missouri Utilities Co. in Cape and had been a physics major himself. They got there before dark so Trickey could become familiar with the terrain, and by about 8 o'clock they were flying 4000 feet above Clearwater Lake. Lights in the towns below were blinking on, but the stars had not yet come out. The sky was a dark blue.

Rutledge was scanning across the top of the plane's windshield with binoculars, he says, when a pulsating off-white light appeared.

"I said, 'Hey, I think we've got a friend up there!' " he recalls. "I wasn't terribly excited, but then a second light came on, also pulsating. They seemed to take turns pulsating.

"Then the bottom of my world fell out for a moment. Seven more lights switched on!"

Rutledge says that this was "interesting for a physicist." According to his account, there was a shimmering haze around the first three lights in the line, which seemed to bend away from the plane. Tiny beads of light appeared to be shooting out from their centers.

"Obviously, I was screaming by this time. It looked very odd. Trickey shouted, 'Where? Where?' and I pointed. We increased speed to 140 miles and changed direction by 20 degrees. I continued to observe for a moment and then handed the binoculars to Kambitch, but the lights were gone.

"It was the greatest emotional experience I've ever had. It really impressed me a great deal. After that, I became — well, not a believer — which is a term I don't like — but an accepter. I thought we were surely into something strange."

A couple of weeks later, Rutledge brought an investigative team to Piedmont for two weeks. Shortly after they arrived, he heard that an odd amber light had been seen near Farmington, and he and Wilson drove up the next evening. He describes making several unusual sightings there, including one of an extremely large object that passed over him and several other observers at the Farmington airport. By his account, as the object moved away he could see four lights — red, white, white, red — across its rear and even pick up in his binoculars what appeared to be ribbed glass on one of the lights.

At this point, Rutledge enlisted members of the Southeast Missouri Astronomy Club in his investigation. For the next year, with their help, he spent almost all of his spare time trying to find out more about UFOs. In the first seven months, he says, there were only six nights on which he didn't work on the project.

He spent hours standing along country roads south of Cape Girardeau, swatting mosquitoes in the summer and trying to keep warm in the winter, constantly studying the night sky through binoculars. Usually, one or more members of the astronomy club were with him.

"Mercury's out. We should be able to see it."

"You can see Mercury with the naked eye now?"

"Yep."

It was really cold now, and everyone was walking up and down the road to keep warm. Mark Rutledge had given up on peering through the 800mm lens and was concentrating on shivering.

"Mark," said Jim Hickam with a grin, "if things ever get rough at your house, you can always get your dad for

See UFOs, Page 11

index

UFOs

■ FROM PAGE ONE

crewcut, is an elementary school science teacher. Like the other astronomy club members who were there, he didn't seem like the kind of guy who would spend his time looking for UFOs. But then, neither does Rutledge.

"A person has to be dedicated or a damn fool to stand out here on a night like this," said Hickam, grinning again. "They've had their hands freeze to the telescope. Bob Adams had a heck of a time getting his hand off one time.

"In the summertime, it's different. There's no lack of activity; you just start slapping. You don't even have to aim. Anywhere you hit, there's gonna be mosquitoes."

The president of the asronomy club is Robert Adams. He farms south of Cape, has his own observatory and is a photography buff, too. He also has a high regard for Rutledge.

"When he got involved in all this, Dr. Rutledge really didn't know all that much about astronomy or photography," Adams said, "but he made sure that he learned. That's the kind of guy he is."

"I don't really want to accept this myself," said Rutledge. "I know my subconscious doesn't. At times, I've tried to drift away from it, but then I look at my pictures and. . . ."

After the exhausting first year, a period when there were a great many UFO reports to chase down, Rutledge has cut back somewhat, concentrating on putting together his data and gathering equipment for another splurge of reported sightings. His major concern right now is finding a van in which the equipment he has accumulated can be taken into the field without spending hours loading it in and out of a pickup. He still scans the sky most evenings, though, often on a walk with his wife.

Of the more than 140 UFOs he believes that he has seen, 20 or 30 were sighted from near his home, which is in a subdivision that sits high on a hill in north Cape Girardeau. He says that one of them was a daytime sighting of a large,

Continued from previous page

ST. LOUIS POST-DISPATCH Sun., December 4, 1977 1 11.

bullet-shaped object that appeared to be several miles away. On another occasion, he says, he saw a disc-shaped object from his office window.

He knows that reports like this are not easy for most people to accept.

"It took all the courage I've got to do this," he said. "Sometimes I marvel that I have been able to stand up in faculty meetings and say these things.

"People ask why I have seen so many. Maybe it's because I look. I've often wondered why I saw something the first two times I went to Piedmont. If I hadn't, I probably never would have gone back.

"If I publish the scientific paper that I'm working on, I'm going to be attacked because of it, and I'm just going to have to weather that. After all those nights in the field, though, I get a little disturbed when I'm criticized by scientists who have never left their offices."

Although he is reluctant to discuss his theories about UFOs, he does say that he thinks they are intelligently controlled and that they react to being observed.

So what if they are and he happens to meet some of the folks on board?

A smile appeared on Rutledge's face.

"If that happens," he said, "I'll quit and drink a lot!"

Article:

Professor Reports on Strange Lights

Neosho Daily News

ST. LOUIS (AP) — Missouri physicists scratched their heads Saturday afternoon as a Southeast Missouri State University professor presented a study of recent sightings of strange objects in Missouri skies.

"Hmmm," was the collective reaction of the Missouri section of the American Association of Physics Teachers meeting at Meramec Community College here.

"I don't want to refer to them as UFO's or flying saucers. They were unusual lights. I don't want to speculate beyond that," said Dr. Harvey D. Rutledge, head of the physics department of the university in Cape Girardeau.

Dr. Rutledge and a group of physics students and astronomers conducted field observations for 78 nights in the cities of Piedmont, Farmington, Fredericktown, Cape Girardeau and Sikeston. He said nothing was observed on 40 of the nights.

He said that during the other 38 observations, which covered the period from April 6 to Nov. 10, "unusual lights" were observed and photographed.

Rutledge presented about 20 slides, most of them time exposures of six to 10 seconds, showing streaks of light in the darkened sky. One photograph showed a wiggly streak, which when enlarged, revealed coil-like spirals originating from the streak.

Color slides showed that the lights were usually orange and green.

Rutledge said some of the lights resembled "charged particles in a magnetic field."

During a question-answer period, one physicist speculated that such a pattern could have been produced by a microwave, such as that used to transmit television pictures over long distances.

Rutledge agreed, but said scientific instruments were not available during his study to investigate such a possibility.

His investigation was initiated after Reggie Bone, Clearwater High School basketball coach, and five of his players, reported watching an unidentified flying object Feb. 21. Since that time there have been numerous reports of such sightings in the area and across the Mississippi River in Illinois.

"I told them I wouldn't touch that with a 10-foot pole," Rutledge said as he recounted how the study began. "But then, people started getting shook up down there."

The mustachioed professor, who has been at Southeast Missouri State for 10 years, said he was "shocked" by what he saw the first night of the observations.

He also reported during the 1½ hour presentation that on two occasions he observed airplanes following the lights and added that similar reports had been made in the Piedmont area as long as eight years ago.

One man, who said he was a physical scientist for the Shell Oil Co., said he observed similar lights in the St. Louis area Nov. 1.

"Was that at about 10:20?" another physicist asked.

"Yes it was," replied the Shell worker.

"I saw it too," said the physicist.

Chapter 5

1896 to 2019
Wayne and Reynold Counties, Missouri

By Debbie Ziegelmeyer

MUFON Case Reports

CASE #103845
October 6, 2019, 1:30am
Piedmont
"Viewed total of five spears, and a bright beam moving erratically in all different directions. One spear was inside of a very vivid circle."

CASE #97818
June 15, 1971
Mill Spring
My friend, my 3yr old daughter, and I, were curious about the lights darting in the distance above a meadow by Brushy Creek just outside of Mill Spring, MO. It was between 11:00 and 12:00pm (midnight) and a full moon across the meadow just to the left of the county highway.
We were in a Volkswagen Beetle and parked facing the field. We were watching the UFOs speed, dart and almost play in the field. Suddenly quite shockingly, one rose out of the field right in front of us about 20 ft. away. It hovered for a second or two, then came toward our car. We were terrified and tried to back up and get away quickly. It continued to hover over the car. The light was very, very intense, and white. We drove back toward Mill Spring as fast as we could. My daughter was crying and just as quickly as it came out of nowhere, it left straight up and back out to the field.

Again, about a week or two later, I was driving across Clearwater Dam Road just outside of Piedmont, MO. I could not believe my eyes, but I saw a huge glowing, round object at the bottom of the dam. It was underwater and it emitted a greenish glow, probably because of the water. It was late at night, and I was returning home from work. I clearly saw an enormous UFO with what appeared to be a double row of small windows all the way around it.

It was glowing so bright that I could see it clearly. I was terrified that it would come out of the water and chase me. That time of night, there were no people around. I felt cold and almost disoriented as I tried to drive back home. I will never forget it.

Again, my friends and their mother were driving back to Cape Girardeau, MO at night when out in a corn field to our left, I saw a UFO out in the field. It was round with two rows of very small glowing windows. It was just sitting there about 7 ft. off the road.

CASE #93578
July 22, 2018, 9:30pm
Piedmont

The witness took his dog out for a walk. While standing in the yard, he looked up and saw a star moving downward, stop, zig zag sideways, move up, and then down. Another light close to the first to the left had a white light coming from a blimp like shape with a dark center. The witness watched until both disappeared which was about 30 minutes.

CASE #93253
June 29, 2017, 1:00am
Richmond

"I was in Piedmont the week before July 4, 2017, visiting family. My wife and me were staying in an RV on my parents' property. At about 1:00 or 2:00am in morning we walked out of my parent's house and headed to the RV. I noticed about two hundred yards to the east of my parent's house was a yellow orange light about 20 to 30 feet high just above tree line. I told my wife I was going to walk down and check it out. I walked down a gravel road towards the light. It would only let me get about thirty feet away from it and it would back off leaving a buffer zone. I could hear wild hogs going crazy in the woods to the north of me, so I decided to get my car and my wife. When I made it back to RV my wife was already in bed and would not get up. I could only get so close, so I was determined to sit there and watch it till it left which was over 4 to 5 hours. An hour before daybreak it started to vibrate and change shape. I knew something was getting ready to happen and I wanted to share it with my wife, so I drove back to the RV to wake her up. I got her to get up and walk out of the RV and look at the object, but she turned around and said she was going back to bed. It was getting light, so I just walked back down towards the object. When I got back it had centered itself in an opening in the valley and had turned bright blue. The light was so bright it looked like points of light were coming off both sides.

As I watched it the point on the right moved downwards towards the other point. When it did this, a black cloud formed around it, and the point that moved to the bottom now moved back in the same direction but pointed straight up. The other point moved down, then the craft made a loud whizzing sound, sucked up the black smoke, and lifted it. This was about twenty feet away from me. I'm not a religious man but one of the first things that popped into my head was it looked like an angel with wings being the point of light. It only lifted about twenty feet and repeated the same process as before to create the cloud, it repeated this process about fifteen times till it just rose up high enough and disappeared. When it disappeared, what looked like big translucent rain drops and fibers, rained down to the ground right under where the craft was hovering. Before the drops and strings hit the ground they curved and came right at me. The drops were absorbing into my skin. When I tried to wipe them off my arms with my hands there was nothing there. I didn't have a cell phone because you get no service out in this area.

CASE #90234

June 1963 or 1964, 8:00pm

Lesterville

Two cousins eleven or twelve years old, were in the back seat of a car with their aunt and mother, one of which was driving, one was a passenger. The incident occurred in June 1963 or 1964 as they were traveling down highway 21 heading toward Lesterville, Missouri. The sky was "pitch black" and they were the only car on the road that they could see. There were no other visible lights until a large bright light appeared from behind them. They could see out of the back of the car, a very bright light coming toward them until they were engulfed in it. They could not see any shape only bright lights, and everyone in the car was totally silent. "It was like we were floating and in a trance of some sort. No one was talking or screaming, and no one was afraid." This reaction did not make sense to the witnesses. Suddenly, right next to the car, close enough that they felt like they could reach out and touch them, there were two Greys which they could only see from their side view from the shoulder up. Both Greys turned and looked at the witnesses in unison but still, they were not afraid. Everything was totally silent.

The witnesses did not know how long they were looking at the Grays, or how long the Grays were looking at them because they seemed to be in limbo. The car was no longer moving.

After an undetermined amount of time, the Greys turned in unison moving back farther away from the car. The witnesses seemed to think that they were looking out the front of something.

"We had to be up in the air or something because otherwise we couldn't have been at the same level as the Greys were. We were in a car on the ground, so it didn't make sense. Either the whole car was lifted up or just we were. I just know we felt like we were floating."

At some point the car was traveling down the road again. None of the passengers were speaking about what had just occurred, and no one acted like anything had happened. Then, the witnesses saw the bright light move in front of their car. It was round shaped, and very bright. It ascended at a great amount of speed, straight in front of the car, then moved to the side and then another angle and finally moved straight up and disappeared.

The witnesses had no idea of the time lapse, if they were in the same location they began at before the incident had happened, or if they were closer to their destination which was their grandmother's house. Nobody spoke about the incident, no one seemed to be afraid of what had just occurred, it was like it never happened. A couple of days later, the four of them, the two adults and two girls, began to slowly talk about the incident a little bit at a time. One of the aunts drew pictures of what they had seen. The pictures that the aunt drew looked exactly like what we now describe as a Grey. The witness described the being as follows; "It had a whitish grey smooth type of skin with big, slanted eyes barely any nose and very small mouth no ears. I don't remember the expression on the Grey's face ever changing." The family spoke of the incident periodically over the years, but the family members who did not experience the incident were thought to never really believe that it had occurred.

F-18 jet Photo: Wikimedia.com Public Domain

CASE #87433

October 18, 2007, 9:04pm

Bunker

"My husband called me to go outside because there was a large amount of air force jets flying over around us. No jets ever fly at night in our area of the woods. We were both outside watching these jets encircling something, then we saw this massive light. It was bright with

orange and blue on it, then, it disappeared. A few minutes later it looked as if jets were chasing something, we saw this gigantic, huge bright light and seconds later it went away!"

CASE #74186

January 28, 2016, 2:00am

Piedmont

The witness owned a bar in Piedmont and had just come home from closing for the night. He decided to go outside for some fresh air when something caught his eye. He began watching events such as multiple discs across the sky from viewpoint. He also saw one orb fly circles around a jet as if it were playing with it, only to veer off in a completely different direction almost a right angle. Two craft were similar in size and very visible with the moon as a backdrop. He watched many different crafts but the one closest to him had red pulsing lights that descended towards Clearwater Lake just behind his house. The craft emerged moments later rising in a different path and then descended. A triangle shaped craft was also observed but other than just being in the vicinity, it basically just hovered within his view all night. There were multiple aircrafts with red flashing lights on their wing tips which were also seen that same night. The witness, having just moved into the area six months prior, became a little spooked and went indoors to watch. From the windows, he saw one yellowish lighted object fly above the tree lines. He tried to capture some of these craft and lights with his cellphone but could not figure out why the camera would not work. He watched the activity in the sky overhead until sunrise, when it became too bright to see anymore so he went inside to bed. The next night he began watching the sky again during the afternoon and noticed four jets in the air flying a grid pattern in the area. On his drive home late that evening along highway AA, there is a curve where the speed is reduced to 15mph. At the curve there is a gravel road offshoot where he saw a two-tone red and pinkish craft. It had a light at one end long and flat not red, the bottom was pink, and the top was long and slender. It was hovering but was not parallel to the surface, so he stopped at the mouth of the gravel road and got one picture which he showed to me when I met him for an interview. DZ-SD

CASE #68527
July 15, 1974, 10:00pm
Piedmont

"Me, Mom, Dad, an uncle, an aunt, and some cousins were gathered around a BBQ fire pit at the Clearwater Lake on a Saturday night. Dad comments that it is about time for sputnik to come through the sky. Shortly later, Dad proclaims, "there's Sputnik" and sure enough, a satellite was heading south to north and was nearing noon high directly above. Only a moment into watching, Dad says here comes another one and another one and another one. Three objects that looked equally bright and equal in size to the satellite came zooming and made a triangular formation around the satellite and synchronized their pace with it. Each object came from different directions at un-comprehendible speeds given satellites travel in the range of 17,000 miles per hour in orbit. One came from what seemed like due south, the next was from the west and the last was from the east. They showed up one at a time, but immediately after each other. After watching the three objects track the satellite for only seconds, all three objects zoomed off back in the directions they came. Unlike their arrival, the departure was exactly simultaneous and moved even faster than when they arrived. There was no sign of ramping up to speed, their acceleration seemed instantaneous."

CASE #65498
May 15, 1974
Park Hills

"In the Spring of 1974, a silver disc caught my eye through the windshield just as it moved ahead of me over the road in Piedmont, Missouri. I didn't know if it had been following me. I stopped the car, got out and looked up and it had stopped moving. It was silver with rings on the bottom similar to the rings on a vinyl record or to explain better, rings in an aluminum pan. There were no lights except the streetlights which seemed to be reflecting on it. There was no sound, and no windows. It was at least as wide as the two-lane road. I looked around and there was no one around, not one car or person that I could see. There was a tavern on the right, and it was closed. It was stopped in the road just outside my car door. It slowly started moving away from me after crossing the road and it then slowly disappeared from my view. I could then see that the top was the same as the bottom."

CASE #58048
July 12, 2014, 9:45pm
Piedmont

"It was about 9:45pm, and I was standing with my friend facing north on site C18 at Bluff View campgrounds on Clearwater Lake, when out of the corner of my eye I saw an orange glow to the west. It was completely silent moving east with a slight northerly directional change near the end of the event. My wife who was heading back from the shower nearby, and neighbor on C19 to the north, were also able to get a look at it. About midway thru the sighting, I looked at it with my binoculars. It had a bright yellow glow in the center like a 60w light bulb surrounded by a red/orange blur. A few seconds before we lost sight, the object moved erratically then it was out of sight. The event lasted no more than three minutes from my perspective. I knew there were sightings around this lake in the 1970s from locals telling me stories when I was a kid. So, I brought the field binoculars just in case we saw something. I have been looking up for years and finally I caught something I can honestly say was unidentifiable. Ironically, it is in the place where I was first turned on to UFOs and basic astronomy. This was definitely not an airplane or helicopter. I know enough about known aircraft to prevent a misidentification."

CASE #47649
May 24, 2013, 11:10pm
Piedmont

I heard a strange humming sound outside of my house which is on one-hundred acres of wooded area near Piedmont, Missouri. I went outside to investigate what the low humming was and saw a large brightly glowing orange/red ball of light on the NE corner of my house, just above the roof line and approximately fifteen feet from the house. It looked like a solid object underneath a halo of light surrounding it. I was standing about thirty feet away from the object. I was in shock and stood there, frozen, while calling to my 36-year-old son to come out to see it, which he did. We both watched this object for about twenty minutes or so as it hovered in the same spot, wavering, then it moved to the southeast. I noticed that there were at least ten more of these objects in the woods at and below the tree line about one hundred yards from our position. The object moved very slowly towards the woods, seeming to be unaware of our presence or not to care that we saw it. I tried desperately to get my cell phone to come on so I could take a picture, but it would not work. I thought the battery was drained, but after the object was out of sight the phone resumed working.

Next, we turned to go back into the house, and something caught my eye. There was a four foot tall humanoid being standing next to the tree and a little behind it. It looked greyish but it was hard to tell in the darkness though there was a full moon out. I just got a glimpse of it before it vanished. Oddly, I did not react by yelling, or running away. The impact of what I saw did not hit me until the next day. My son said that he did not see it. When we got back inside, I was shocked to see that it was at least 50 minutes to an hour later than when we went outside, as it seemed like only 20 minutes of time had lapsed. Next, we heard a helicopter approach and looked out the window as it flew in the same direction as the UFOs. I have never seen a helicopter out here in my entire life. We did not go outside to look at it.

The fact that both of us were able to go to sleep after that was amazing to say the least. It was as if we were in some kind of a trance-state. The next evening my son was in town getting some groceries when he was approached by two policemen. They asked my son how he was doing and if he believed in aliens or ever saw an alien. He replied that he did not. They again asked him the same questions, which he thought was very odd and again replied that he did not. There had been no prior conversation, and he does not know these men very well so thought it very odd that they would say something like that. He was afraid to tell them what he saw the night before. This event triggered the memory of the prior night and when he returned home, he asked me if I recalled the same things he did, which I did recall, including the sighting of the being by the tree. We both found it extremely odd that we did not think of the previous night's event earlier in the day, or that we were not more upset by it, and had not called anyone about it. How did the police know about this event? We spoke to no one about it.

CASE #41183
July 26, 2012, 3:50am
Silva

"I was driving south on highway 67 north of Silva, Missouri. I saw a blueish light flashing in the distance. I assumed it was a radio tower. As I got closer, the light got brighter. I looked over to the east and above the few houses in the area and I could see 3 large square or rectangular blue lights (the color of HLD headlights) that flashed in a pattern that seemed to chase its way around whatever they were attached to. I could only see three lights, but I could see the lights were flashing 360 degrees from the source. I couldn't see what the lights were attached to. I didn't hear any noise and the lights didn't move. They were just hovering, frozen in one spot. I wanted to stop and take a picture, but I didn't feel comfortable stopping and there were no other cars around. I watched the lights in my mirrors and after about a mile I saw the lights rise above the trees and disappear."

CASE #32802
October 22, 2011, 10:45pm
Patterson

"I noticed a slow fire orange object or light moving west to east. It went out, disappeared, and had no sound. Six additional identical objects came down the same path several minutes apart, all with no sounds, all separate objects, and blinked out at same place in sky. These light objects were 500ft above tree level, and very intense orange in color. I was on my front porch which faces north when I saw the lights."

CASE #26420
November 6, 2010
Black Township

"I was enjoying a beautiful night under the stars, noticed two weird streaks of light within a twenty-minute period approximately 8pm Saturday night that were very fast moving. First one was in the west sky and the other one in the east. The first one I thought must have been a shooting star which was visible to me for approximately 10 seconds, shot across the sky with a white light and white trail and then vanished. When I noticed it again later in a different spot, I thought that was quite a coincidence."

CASE #20222
June 20, 1980, 9:00pm
Piedmont

"My wife and I were driving down the entrance to a campground on Clearwater Lake about nine thirty pm when a glowing red sphere came out of the woods about 150 feet ahead of me and headed toward our car. I was going about thirty mph and the object was coming toward me at about the same speed following a slightly wavy path. It came within a couple of feet of the hood of my car and then raised up a little and flew over the driver's side fender and right past my open window about thirty inches from my face. I could have reached out and touched it. It was about the size of a soccer ball and perfectly spherical and glowed red the color of the brake lights on a car. I looked back after it passed and watched it follow a wavy path down the road behind me until it went around a curve and wasn't visible anymore. We just stopped the car and sat there in awe of what we'd both seen."

CASE #6243

December 27, 2006, 5:25pm

Piedmont

"Between Christmas and New Year's, me and my 12-year-old daughter saw an orange glowing UFO. I was coming home from the auto repair shop when I turned down the road I live on and something caught my attention, a bright light. When I turned and looked to the left there was a very bright orange light in the sky to the west-south-west. It was not moving, was low in the sky, glowed bright then slowly dimmed till it went out. The light came on in a different place." As described by the witness; "If you held a wooden ruler at arm's length against the sky it moved that far to the right or west. I didn't see any object during the movement the only way I know it moved was because it glowed in a different place. After it moved it glowed again very

brightly then it dimmed and went out. Then it moved back to the original place and glowed again then it went out. I saw a jet fighter coming from the west southwest from the other side of the UFO object from where I was. It came almost to the place I last seen the glowing object then the jet turned south. I saw what looked like a small silver disk fly up the left side of the jet to the front tip then back down to the end of the jet or rear of the jet. Then it flew up to the front once more then flew away to the north. The jet appeared to try to follow it but could not turn fast enough."

Independent Reports

1986 Sam Baker State Park, Reynolds County

Michael R. and his four-year-old son Michael Jr. were in the park camping for a weekend. Michael Sr. wanted to spend some quality time with his son, so they decided to go to Sam Baker State Park for some swimming, fishing, and camping. Rather than pitch a tent, the father and son were sleeping in the family 1986 Dodge window van. At approximately 1:00am, an extremely bright white light enveloped the inside of the van as they were sleeping. The light was so intense, it woke Michael Sr. who was immediately temporarily blinded. There was no sound, just a bright light that lasted approximately two minutes. When Michael Sr. grabbed the van door handle to open it, the light stopped. He went outside to see if he could determine where this light had originated from, but he could not see anything. There were no vehicles, boats, or people active in the area and no one carrying a bright light. As Michael Sr. looked to the distant sky, he noticed a small white light moving away into the darkness.

Michael Sr. was wearing a t-shirt to bed and his legs were covered with a blanket just below his knees. Michael Jr. was covered to his shoulder line. Both woke up that morning with severe burns on the parts of their bodies not covered. Michael Sr. had severe burns on his arms and the upper part of his legs and Michael Jr. was burned so badly his eyes were swollen shut. Because of the burns, Michael Sr. decided to pack up and go home for treatment. On their way out of the park, they met a park ranger who told them no one else had reported a bright light or had the experience they had.

Both father and son went to the emergency room where they were examined for first sunburn then poison ivy. The burns were not caused by either. A specialist was brought in to examine the father and son and determined that the burns were caused by exposure to radiation. The father and son experienced flu like symptoms for the next few days but as far as they know, there were no permanent effects or scares. Michael Sr.'s wife took a couple of pictures of her husband and son showing the burns. I was given a copy of these which I did not include with this report to protect their privacy. DZ

June 25, 1952

Poplar Bluff American Republic Newspaper

Flying saucers have been spotted near Poplar Bluff again, according to a letter received here today signed by one Frank Harty, Route One, Poplar Bluff. He writes: "After reading the article in Friday's paper about the flying saucers, I thought it might be my duty to report the ones I saw, but I don't know where to report. On June 9th at 7 o'clock that night, I saw two and possibly a third of these objects about 20 degrees north of east and going in a cloud formation. I was able to judge the approximate distance and size" The story Friday reported that J. W. Adam, Jim Freeman, and Charles Feverston of this city caught a glimpse of the object high in the sky in the northeast last Thursday afternoon. They described it as a round, shiny object like a hubcap of an automobile, and that it appeared back and forth in view several times. They said it moved very slowly.

Cattle-Killing UFO of 1896

Fate Magazine

"Our sighting took place, as I recall, in August 1896. My parents Ezra and Serepta Vickers, my brother Ben and I lived on a farm in Howell County and had lots of cattle and horses. We worked hard, and in the evening, when the chores were done and supper was over, we would sit in the yard resting and trying to pick out the different constellations like the Big and Little Dippers. It was a favorite game for Ben and me. The night we saw the UFO, the sky was clear, and a million stars were twinkling. It had been a hot, dry summer and we needed rain. All at once we saw flashing lights. At first, we thought they were a star shower. But the "stars" stayed in a circle as if they were all strung together like beads. Meteors scatter when they fall, so we soon realized we were seeing something else. The circle of lights just kept whirling and falling toward us. In my mind I can still see those bright lights. They were brighter than any star. As small as I was, I remember thinking there was nowhere to run or hide from this thing. What was it going to do to us? For some reason I thought it was unfriendly, that it meant only to do us harm. I ran to my mother and clung to her skirts, crying, and shaking so hard that I could barely stand. I'm sure she was as frightened as I, but she tried to comfort me.

At that time, I had never seen a balloon, airplanes were in the future, and nothing belonged in the sky but the sun, moon, and stars. We watched fascinated. We could not take our eyes off the object as it moved west right toward us, turning, and falling out of the night. Just when we thought it would crash in the yard, it stopped and hovered over the barn. We could now see it was a large saucer like shape. Its lights were blinding. The whole barn lot lit up like day. My father grabbed up the baby who was sleeping on a pallet and we fled into the house. None of us dared voice our fear that God had returned, that this was the end of the world. My father handed the baby to my mother and went into the kitchen to look out the window. The lights were still hovering there. We could see no sign of life, no forms moving out of the object. If there was a door, it was on the other side. My parents gathered us children in the living room, and we huddled together absolutely terrified, waiting for a knock at the door or for the door simply to burst open. My father started to pray, tears running down his cheeks as he hugged all of us. It was one of the few times I ever saw him cry.

I don't know how long we knelt there. All I know is that it seemed like hours. Except for my father's voice murmuring in prayer, there was no sound. The silence was the worst of all. We waited helplessly, not knowing what was going on outside, wondering what might happen to us at any moment. At last, my father got up the courage to go back to the kitchen and look outside. The thing had gone. We slept very little that night. Was the saucer gone for good? Would it come back? Lying wide-awake in my bed, I could hear my parents downstairs as they talked far into the night. Every so often my father would go to the door and look out. And the horror was not over.

The next morning when Ben took his dog Cappie and went out to the pasture to bring up the milk cows, as he always did, he came running back to the house scared out of his wits. Mother and I were getting breakfast when he appeared in the kitchen door so pale, I thought he was going to faint. He kept pointing toward the barn, stammering words that made no sense. Finally, Mother got him seated in a chair and bathed his pale little face with cold water. Then he told us what he saw. As he and Cappie neared the gate that opened into the pasture, Cappie stopped in his tracks, whining and barking. The grass was almost knee-high to the cows, so the animals were not immediately visible to a 10-year-old boy. But in a large patch of burned grass were three of our steers lying dead on the ground. Their bodies were beginning to bloat from the heat; their legs were poking out stiff as boards. Cappie snarled, showed his teeth, and refused to go near them. That was when Ben ran back to the house.

Father, who had come into the kitchen while Ben was talking, ordered us to stay in the house while he went out to the pasture. When he came back, he was as shocked as Ben. He told us he had examined the steers and as far as he could see there was only one thing wrong with them: they had been completely drained of blood. The only marks on them were some dried blood on their throats from two puncture holes in the jugular vein; these looked as if they had been made by a two tined fork. The idea that something or someone had dropped out of the sky to take the blood of animals sickened us. Had they drunk the blood, put it on their food, used it for medicine? What might have happened if they had found us? Would we too have been killed? We were too frightened and nauseated to eat the breakfast sitting on the table in front of us. In those days you didn't waste good food, but we could not eat. Father tried to assure us it was nothing to worry about. He said that probably some wild dogs or wolves had done it. We didn't believe it and we could see he didn't believe it either. We had seen too many animals attacked by dogs and wolves. They always tore the animals apart. They didn't drink their blood.

I was always a nosy child, always eavesdropping when the grown-ups thought I was somewhere else, so I listened to my parents talking about the possible ways the steers could have been killed. Had a neighbor killed them and been frightened away by the lights before he could cut it up? A neighbor would have had to shoot the animals and in that isolated country a shot could be heard half a mile away. If someone had punctured or cut the animals' throats, he first would have had to tie them up. There were no ropes. In any case, animals don't die at once when their throats are cut; the wounded animals would have bawled their heads off. The other cattle would have stampeded and made a terrible racket. We had heard nothing. America was in the middle of a depression that summer and nobody would have left the animals' hide. A cowhide was worth about fifty cents—a day's wages for a man that year. If a neighbor had killed a steer, it would have been to feed his family. These carcasses were just left. Furthermore, a beef has to be hung up in order for the blood to drain, not left lying on the ground. That's what was so strange. How did the aliens, if that's what they were, drain the blood? There was no blood anywhere; every drop was gone. Had this happened only to us, or had they killed other stock in the area?

Later that week, when the newspapers came from St. Louis, we had our answer. There were several stories of just such incidents all over Missouri that night. People had seen the hovering lights and, in each case, only three animals were drained of blood no matter how many cattle were in the pasture. Where the saucers had hovered, grass was burned away by the heat of the object. In our pasture where the ground was burned, nothing ever grew again. The patch was bare. It still is as far as I know. My father never spoke of that night again and he did not let the rest of us speak of it either. A note from Shannon Graham: I am certain that my grandmother Pearl Chenoweth's account of her 1896 UFO experience is an accurate one. At the time she related this story, she was dictating her memoirs for the Missouri Historical Society and was helping people with genealogical research. Everyone who worked with her was amazed at her accurate recall of dates, places, and people. She was alert and active until her death in 1984."

Although not reported in volume in the Brushy Creek area— during the 1970s, numerous cattle mutilations along with UFO sightings were being reported in Elsberry, Missouri 180 miles north of Piedmont and just sixty miles north of St. Louis and in other areas of the state.

Chapter 6

Recent Interviews

By Debbie Ziegelmeyer

Paul

Herculaneum, Missouri
Interview by Debbie Ziegelmeyer

On February 21, 1973, Paul was living in Piedmont and attending Piedmont Junior High. He remembers the incident involving basketball coach Reggie Bone and his students very well. The students were returning from a basketball tournament in Essex Missouri which they unfortunately lost, when they witnessed a large UFO craft hovering low over an open area of nearby woods. Paul was asleep in the first of two school buses following coach Bone's car back from Essex which was about an hour and a half trip. The bus driver noticed coach Bone's car stopped ahead with the coach and the students looking at something off the side of the road. The commotion awoke Paul from sleep. He said that people in the bus were looking out the windows at a strange object with multiple lights. The students were acting confused because most of them had never heard of a UFO. Some of the students, maybe a dozen boys, got out of the bus and ran across the field placing themselves directly under the craft. The craft slowly moved towards the woods, took altitude, and moved off into the distance.

So many people from town were seeing strange unidentified objects in the sky that they began gathering at the dump on weekends to sky watch. The dump was the highest location in the area. Paul remembers being at the dump with his family and seeing lines of parked cars belonging to people who had come to sky watch. The location of the dump was eventually sold, and the dump moved to a new location. The hill is now grass covered with a cell phone tower in the middle. The towns people still felt the need to have a lookout, so a roadside park called Scenic View was built which is about ¼ of a mile from where the dump was located.

Paul remembers being at the drive-in movie which

is located on the edge of town when other mass UFO sightings occurred. He said that he can remember on more than one occasion while watching the movie, someone would yell "they're back, turn off the projector!" The projectionist would turn off the movie and park lights and they would watch strange lights dancing around in the sky above.

Paul knew Jean Coleman and Kathy Leech who reported the sighting over Clearwater Lake. He said Piedmont is a small town and everyone knows or knows of everyone else. Paul's father was an air traffic controller back in the 1970s and told him he wasn't sure what people were seeing but a lot of people were watching something.

My personal family sighting

April 1979 2:30pm
Jefferson County, Missouri

In April of 1979 my family and I were in Arnold, Missouri shopping. There was a Plaza located at Highway 141 and Highway 61-67 with a K-mart and adjoining grocery store. I went into the grocery store to shop while my husband Wayne took our seven-year-old son Wayne Jr. and 3-year-old daughter Nikki with him into K-mart. The sky was clear but windy that day to the point where my long hair was whipping back and forth across my face.

I finished grocery shopping at about 2:30pm and began loading my groceries into the back of our pick-up truck when I noticed something very unusual in the sky above. In the eastern sky at about 60 degrees were two silver metallic spheres one directly over the other with just enough space in-between to notice separation. The objects were stationary in the sky not moving or shaking which seemed very unusual especially with the wind blowing at least twenty mph. The objects were perfectly round with defined edges, no markings, just smooth silver metallic in color each being approximately 1/5 the size of the moon in the sky during the day. I watched them for about ten minutes while standing next to the back of our pickup.

I was waiting for my husband Wayne to return with our children not wanting to leave my groceries exposed in the back of our truck for someone to steal.

2 When Wayne returned with our children, he asked me what I was looking at and I pointed up. By then, several people in the parking lot were looking at the same strange objects stationary in the clear blue sky while the wind blew fiercely. The four of us got into the truck and drove south on Highway 55 towards our home in Barnhart. As Wayne was driving, he kept asking me and our children if we could still see the objects from the window which we could. At the time of the sighting, we were living in a subdivision named Parkton which is located in Barnhart 8.5 miles from K-mart. What was very unusual was the fact that we could plainly see the objects in the sky as we traveled down Highway 55 south towards home. This would mean that these objects had to be at a very high altitude and from their size, very large.

When we pulled onto Catlin Drive where we lived, most of the people on our street were standing in their driveways looking up at these two objects. I asked the neighbors if anyone had reported seeing these objects and if they knew what they were and was told by several of the people outside that the police were called, Scott AFB was called and the news media. We all watched the objects which were still stationary in the sky for about twenty more minutes when suddenly a strange fog rolled in and covered the sky above. At that point with nothing more to see, we all went inside. That night we anxiously watched channel 2 news to see if there was any mention of this.

The newscaster at the time was Larry Connors who

Forest Crawford

Ballwin, Missouri
Interview
by Debbie Ziegelmeyer

Forest Crawford is a well-known UFO investigator. Knowing that Forest Crawford had lived in Missouri in the 1970s, I asked him if he had experienced anything UFO related during this UFO Flap. The following three sightings, two secondary and one personal account were told to me by Forest.

In the late 1970s Forest was supposed to go to Piedmont with J. Allan Hynek and Walt Andrus but had to work. Hynek and Andrus were in the Piedmont area driving down a country road to talk to some witnesses that had found footprints on the ground that could have been made by aliens. As they were driving, they came upon some lights in the sky down the road in front of them, maybe 200 feet above the road. They described seeing three orb like lights in a triangular formation. These were three separate lights which were not attached to anything, so close they lit up the road below. Hynek and Andrus stopped their car and got out to look but since the lights were stationary, decided to get back in the car and drive closer. As they got directly under the three lights the lights split apart and departed at a high rate of speed in three different directions. The light that was the pinpoint of the triangle shot straight up in the sky and disappeared.

Forest's personal sighting occurred in 1977 or 1978 while he and his three high school buddies decided to go on their "first annual canoe trip" which turned out to be the only one. Forest purposely chose the Black River along Clearwater dam because of the history of UFO activity in the area. As they were putting their canoes in the river, Forest noticed a man "checking them out". He asked this man who worked at the dam, if he had ever seen anything strange in the area. The man replied, "oh yeah, we see strange lights all the time". The man told Forest that he sat up on Clearwater dam one night and watched lights come in and out of the water. The man said the lights would come out of the sky and go "plunk" right into the lake. He described seeing lights under the water for several minutes, then they would disappear. About a half hour to an hour later two lights came out of the water and flew away. This was seen by the man on a couple of different occasions.

Forest and his three friends put in the river below the dam and canoed for two days. On the first night they set up camp on the edge of the river downstream halfway between Piedmont and Poplar Bluff to sky watch. Two of his friends were very skeptical about UFOs but the third had just a bit of knowledge and interest in the topic. Forest was just getting into photography and hoping to catch something unknown in the sky, so he came on the trip very well prepared. He had a dual camera mount on a tripod, with two 35mm cameras one loaded with high-speed infrared film, the other with high-speed black and white. The infrared film was slide film and the black and white film was negative. The cameras were mounted side by side so he could fire them up at the same time quickly if he saw something unusual. 2

While camped that first night, the four young men saw three different objects. The first was a starlike object that streaked into their view and stopped directly overhead. As they watched the object, it did a maneuver that he could only describe as a "scribble." The object looked as if someone had a penlight and then began moving it around the sky and scribbling with it. At arm's length it covered about an inch of the sky. It would move erratically around the sky then stop, begin moving again and then stop. The object maneuvered in that manor three or four times then shot off at a high rate of speed in the same direction it came from. The second object was blue starlike which moved across the sky at a high rate of speed and then blinked out. The third object was a diamond shape configuration of four crimson red lights which moved across the sky from the same direction as the previous two objects, due south. The object at arm's length was ½ inch wide and ½ inch across and as it moved across the sky, blocked out the stars in the space between the lights. Forest was able to take four frames of the object with both cameras before it moved out of view.

At the time of the incident, Forest lived in Belleville, Illinois. There was a local photography shop in the area where he purchased his film and regularly took his slides and film to be developed. Forest had met the owner of the store on several occasions and was on a first name basis with Dennis, the behind the counter clerk. Dennis was about the same age as Forest and the two considered themselves to be "buddies." They spoke of the strangeness of Forest's choice of film, the infrared slide film and naturally, Forest was excited to tell Dennis of his up-and-coming canoe trip with his buddies and what he hoped to catch in the area on film.

Dennis was very aware of Forest's interest in UFOs because the two had spoken about the topic in Forest's past visits to the shop. Dennis told Forest he would be very interested to hear about the canoe trip on his return and if he had caught anything on film. Soon after Forest returned from the trip, he went straight to the photography shop very anxious to see what was on the slide and negative film. Some film was processed in house, some not, but previously, slides had always been processed in house. Forest turned in his film from both cameras and returned in a week to retrieve his slides and prints.

The owner of the photography store knew Forest's dad who was a professional photographer. Forest's dad passed when Forest was just two years old but was well known and possibly friends with the photography store owner. Naturally, when Forest would come into the store, he was a bit of a celebrity because of his dad's reputation in the professional photography field. When Forest entered the store to retrieve his prints and slides, the store owner and Dennis began to act strangely, their demeaner had changed drastically. Dennis was very nervous and kept looking over at the store owner and not wanting to give Forest his prints and slides. Forest noticed two men standing at the counter talking to the owner and Dennis looking nervously in their direction. Forest was finally given his slides and film by Dennis and immediately popped open the slide case to see if he had caught the object in frame. The slides were in the box but laying on top were two strips of film with two slides each not cut and entered into slide frames. All of the other slides were processed and framed as normally done.

Forest pulled all of the slides out of the plastic box looking frame by frame and the only frames not properly processed were those four frames of slide film of the unknown object. These frames were not at the beginning or end of the film but taken in the middle section maybe 8 through 11 in the middle of the roll. Forest asked Dennis if he knew why these frames were not processed correctly. Dennis kept insisting to Forest that he could not tell him why these frames were set aside.

Nervously, he told Forest that he did not know why the slides were processed that way because the film was all sent out. Forest asked Dennis why, because slides were always processed in house, but Dennis kept insisting he did not know anything. At the end of the counter standing with the shop owner were two men Forest had never seen before. One was a military colonel with the name Pe-

Forest Crawford, Master of Ceremonies at the Ozark Mountain UFO Conference
Photo: BBC Radio and TV

terson on his uniform. The colonel and the other man listened to the entire conversation between Forest and Dennis then turned and left. Forest asked Dennis if he knew who the man in uniform was, and Dennis replied that he was a colonel from Scott AFB. Forest asked Dennis why this colonel was so interested in him, and Dennis's reply was "I can't tell." Forest tried to speak to the owner of the shop, but he also refused to speak to Forest and from that point on, never did again.

When Forest got home, he looked closer at the removed frames of film. He saw stars, but also noticed streaks across the film. Upon a closer look with a microscope, Forest noticed that the streaks were in fact scratches across the film directly across the point where the object Forest tried to capture would have been. He then took the investigation a step further and plotted what constellations and stars should have been in the field of view on the film in the direction his cameras were pointing. The stars visible on the four film slides did not match the sky in the direction the cameras were pointing that night. The film strips were obviously fabricated. The black and white film did not show or pick up anything unusual.

After some time, Forest returned to the store to ask Dennis if he could tell him what had gone on that day and if he knew who had manipulated his film. Dennis looked around and said that all he could say was that the store owner was friends with the Colonel from Scott AFB, and they both knew what Forest was trying to capture on film. Forest decided it would be best to have his film processed elsewhere in the future and never returned to the photography shop.

Chapter 7

UFOs and Little People Visit Family for Generations

By Margie Kay

In the summer of 1996 a woman named Patricia called after finding my website margiekay.com. I filed this as MJFON case #76419. This occurred years prior to my becoming an investigator for MUFON when I was working on investigations on my own. Patricia was 96 years old at the time she called and seemed to be very sharp. She said that she lived near Piedmont and had a Piedmont address but would not give me the exact location. According to Patricia:

"I live outside Piedmont, Missouri on a farm that has been in the family since the 1850's. I lived there with my mother, father, grandmother, and siblings. When I was a kid, whoever was the youngest had to sleep on the back sleeping porch of the house. None of us wanted to sleep there because the "Little people" would visit whoever was on that porch at night, and we were all afraid of being visited or taken. I remember that my mother and grandmother told stories about the little grey people who visited the farm at night and that the stories had been passed on down through family members since before the Civil War. Some in the family called them fairies.

One night in particular, when I was 10 years old, I was lying in my bed on the porch, and I saw a three foot tall grey being with big eyes walk up to the window and look inside at me. I was frozen in fear. He just kept looking at me. The next thing I knew, there were three of them, all about three or four feet tall with no hair, big black eyes, small noses and mouths, and they were very thin. The three beings were standing next to my bed, and I remember thinking that I had no idea how they got inside through the locked door. I was even more afraid of what they would do to me and wanted to pull the covers up over my face but couldn't. I could not move. I don't know what happened after that. The next morning, I woke up and they were gone. They

Mysterious light Source: Adobestock.com

never appeared during the day, only at night. I never saw a craft of any type, but sometimes late at night we would hear a humming sound coming from the low depression area around the pond about 100 feet from the house, but never went to look to see what it was. I think we were afraid to. These visits were fairly frequent.

We also often saw small white balls of light in the yard and around the farm at dusk but never knew what they were. My cousins and I would hold our hands out and the lights would land in our hands and let us play with them. They would be hovering just above our palms and if we moved our hand up or down, the light would stay in the same position. We could also toss them back and forth between us. The lights were about the size of a softball and had a soft glow to them and were not bright. As kids, we just thought they were something normal and didn't question it. Apparently, our parents either were not interested in these lights or didn't believe us when we talked about them because I don't remember any adults going outside to play with the lights.

The whole family was sure to be inside after dark and we just didn't play outside after dark - the older family members made sure we came inside, and as I think back on this it must have been because of the little people who would "take" us. That is the term my mother and grandmother used - they would say that the little people would take us at night but always brought us back before morning. It was almost like talking about fairies or something magical that was our little secret. It was not something we spoke about to other people - we kept it to ourselves and still do to this day. I still see them sometimes, only inside the house in my bedroom where I sleep now. I see them come in late at night, usually around 3:00 a.m. but for some reason I just fall asleep and don't remember any interaction with them. I don't let anyone sleep on the porch, though.

My children and grandchildren have also seen the little grey people, but they would not like it if I told them I called anyone about this as they also want to keep it secret. We are not so afraid of

them anymore, just consider them as being visitors rather than something that will harm us because no one in the family has ever been hurt by the visitors. I don't want any publicity or anyone to know who I am because they'll think I'm nuts, but I'm not. I may be old, but I am strong and take care of myself, and have all my faculties."
I called Patricia a year later and she told me about another incident she remembered that occurred when she was 16 years old. She and her 15-year-old cousin, Mary, were sitting on the back porch waiting for their grandmother to finish cooking dinner about an hour before sunset when her cousin suddenly got up and said, "I have to go into the woods." Patricia told Mary that she should not go, because it was almost sunset, and they should go inside soon. Mary ignored her and started walking very quickly into the woods. Patricia followed and kept asking Mary to come back to the house with her, but Mary did not respond. It was as if she were on some type of mission and headed deep into the woods.

Finally, Patricia headed back home to get her mother and when she arrived her mother was on the porch waiting for her. Her mother asked where the heck Mary was, and Patricia tried to explain. They waited as the sun went down for about an hour and were just getting ready to call the Sheriff when Mary came out of the woods, disheveled, and not speaking. She was in some type of trance-like state. Patricia and her mother took Mary to her bed and she slept until mid-afternoon the next day. Upon waking, she had no memory of the evening prior and thought it was morning.

A few weeks later, while eating breakfast, Mary looked at Patricia and said, "Why didn't you come help me when I was in the woods?" Patricia asked if she was talking about the time she went into the woods alone, and she said yes, and that she screamed for help because there was a ship in the woods and the grey people took her on board and wouldn't let her go for some time. The family was dumbfounded hearing this, and Patricia did not recall hearing any screaming. She explained that she had asked Mary to come back but she just kept walking without responding. Mary had no recollection of this at all. Once again, the parents admonished the children never to speak of these things with their friends, teachers, or anyone else.

This is one of the strangest cases I've ever worked on, and I wish I knew the location and family involved so I could discuss it with them and see what has happened since. Chances area, the same things are happening on that farm.

Chapter 8

A Case of High Strangeness and Tragedy

By Margie Kay

In late 2009 I was contacted by a woman named Carly (not her real name) who lived on over 100 acres just outside Patterson, Missouri. She told me about events that had begun a few months prior, and she decided to seek out an investigator and found my website. I asked her to file reports at www.mufon.com, and she did so only twice, but she also kept in contact with me by email and phone for several years. This is the first time I've gone public with this information.

October 2011:

On October 22, 2011, at approximately 10:45 pm the witness was having a cigarette on her front porch, and was watching a neighbor's bonfire to her west. She was looking NNW and saw a very bright orange light just above the trees on the side of the mountain/hill. The light was a "hunter" orange color. She watched the object until it disappeared, then ran inside to get her video camera and brought it back outside. She saw another object of the same size and color, moving in the same slow manner over the trees in the same direction and it also disappeared in the same location as the first object. Then four more objects did the exact same thing (6 total) as she watched all of them for a total of 45 minutes. They all disappeared from view in the same location. The distance was approximately 500- 1000 feet away from her, across the road. It would take a baseball held at arm's length to cover one of the objects up. She said that she blinked her flashlight at the last object, and it went out, but she does not know if it did that in reaction to the flashlight or not. There was a plane in the sky in the background while one of the objects was moving by. She is perplexed about how the objects looked orange with the naked eye but when the film is played back on her TV the lights look white. All the objects were round when zoomed in, but more elongated when zoomed back. The objects moved very slowly, there were no emissions, and no electrical effects. She sits on the porch a lot at night and sees unusual things all the time. She said she would mail the original video tape so I could extract the video.

The neighbors who were having the bonfire were yelling and talking about the lights as they went over their heads. She does not know these new neighbors, but she was going to go to another neighbor who does know them and ask him to talk to them and give them my contact information. I asked her to request that they contact me, but no one ever did.

Note: Her son was at his house not far away, and he said he saw an orange light on October 22. She asked him to file a report, but he never did. Piedmont forum/topics is where other reports may be.

Carly lived at the same place for 20 years (as of 2011). She was born in St. Louis, and moved to Annapolis, Missouri in 1989, where she was at her cabin late one night sitting on the porch with her friend, Dr. Harley Rutledge. They saw an army jet, which was very loud, going down the river and past them, when they noticed a light come on that sounded like an acetylene torch, then second and third lights appeared on a very large object the size of multiple jets - then it simply disappeared. The army jet came back, and more jets came into the area, apparently searching for something. The object then appeared again, further down the river. In 2009 she again saw a similar craft with three lights.

In 2010 she took video of a craft coming from the Patterson Community center area (one mile from her house) heading toward her house, but it disappeared when it arrived directly overhead. A second object came from the same place, then disappeared as the first one did. Neither object flue went of sight, they simply disappeared. This gives rise to the speculation that there may be a portal to other dimensions at this location.

Carly mentioned that she has also seen beams of light from the ground going straight up but could find no explanation for them.

Carly has seen UFOs her entire life, and her family members have as well. She told me about an incident involving her aunt and cousins. Her aunt had her three kids with her while driving on a highway one night when her car went dead. She saw a bright light right in front of her on the road. The object left, and the car started again, but they noticed that the vehicle was cold, and it was an hour later than it should have been. The family rarely discusses the incident.

On Dec 12, 2010 Carly got photos of unknown beings with her trail cam she had set up in her back yard.

The following are a few of the many emails I received from Carly:

May 24, 2013

As reported to me by phone: "I heard a strange humming sound outside of my house which is on 100 acres of wooded area near Piedmont, Missouri. I went outside to investigate what the low humming was and saw a large brightly glowing orange/red ball of light on the NE corner of my house, just above the roof line and approximately 15 feet from the house. It looked like a solid object underneath a halo of light surrounding it. I was standing about 30 feet away from the object. I was in shock and stood there, frozen, while calling to my 36-year-old son to come out to see it, which he did. We both watched this object for about 20 minutes or so as it hovered in the same spot, wavering somewhat, then it moved to the Southeast and at that point we noticed that there were at least 10 more of these objects in the woods at and below the tree line about 100 yards from our position.

The object moved very slowly towards the woods, seeming to be unaware of our presence or perhaps not to care that we saw it. I tried desperately to get my cell phone to come on so I could take a picture, but it would not work. I thought the battery was drained, but after the object was out of sight the phone resumed working. Next, we turned to go back into the house, and something caught my eye. There was a 4 foot tall humanoid being standing next to the tree and a little behind it. It looked greyish but it was hard to tell in the darkness even though there was a full moon out. I just got a glimpse of it before it vanished. Oddly, I did not react by yelling, or running away. The impact of what I saw did not hit me until the next day. My son said that he did not see it.

When we got back inside, I was shocked to see that it was at least 50 minutes to an hour later than when we went outside, as it seemed like only 20 minutes of time had lapsed. Next, we heard a helicopter approach and looked out the window as it flew in the same direction as the UFOs. I have never seen a helicopter out here in my entire life. We did not go outside to look at it. The fact that both of us were able to go to sleep after that was amazing to say the least. It was as if we were in some kind of a trance-state.

The next evening my son was in town getting some groceries when he was approached by two policemen. They asked my son how he was doing and if he believed in aliens or ever saw an alien. He replied that he did not. They again asked him the same questions, which he thought was very odd and again replied that he did not. There had been no prior conversation, and he does not know these men very well so thought it very odd that they would say something like that. He was afraid to tell them what he saw the night before. This event triggered the memory of the prior night and when he returned home, he asked me if I recalled the same things he did, which I did recall, including the sighting of the being by the tree. We both found it extremely odd that we did not think of the previous night's event earlier in the day, or that we were not more upset by it, and had not called anyone about it. How did the police know about this event? We spoke to no one about it. I called you to make this report in an effort to see if anyone else has reported anything similar, but I don't want any publicity and don't want to talk to anyone else. The whole thing has me a little scared. Note: We have seen balls of light in this area before, and have seen unidentified objects, but nothing as close or as dramatic as this.

My parents and grandparents owned this same land for many years, and they said that all kinds of weird stuff happened here, but they did not talk about it much. My mom and dad saw a UFO over Clearwater Lake in the early 1970s. I remember that my grandfather talked about some missing time and seeing a red fireball back in the 1940s, and that my parents always checked and locked all the doors and windows in the house at night, even during summer when it was cool enough to leave the windows open. They didn't talk about it much but now that I think about it that was pretty strange, since we are far away from any other houses out here."

Photo of an unknown animal captured by a trail camera. Photo: Carly

Not long after capturing these photos, Carly's trail camera, which is motion –activated, caught some strange beams of light and round lit objects for which she has no explanation.

June 3, 2015

"I think it was the morning of the third, I woke up very early. Then I remembered someone talking to me like a dream! Only I know it wasn't a dream, I was extremely calm as I was listening to this voice telling me things, it lasted for over 2 hours. When I woke up I remembered this happened. By that evening, I came down with this weird burning spot on my tongue and roof of my mouth. Today is the 8th and my tongue still burns. What do you think?"

June 9, 2015

"Had three military helicopters go very low over my home. They were equipped with guns on the front. Came out of the east going west. My partner was in the shower at the time, and it freaked her out. She thought they were landing. Also, I've had two psychic events since this dream. Hadn't even looked at my tongue before your Email. It has like an indentation in the center with smaller lesions going off it! Will try and get a pic".

August 31, 2015

"At approximately 10:00 pm I was walking in my backyard and noticed a bright light in the northern sky. It wasn't a star. I know the sky. I ran in to get my binoculars and went out the front door. As I began to raise the binoculars, still looking at the object a smaller light orb shaped came out of the bigger one, and going in an eastern direction. It was cool to see! Just a week before I witnessed a very fast-moving white light moving towards north from west-southwest. There was an approaching possible jet, the object I saw at first was a bright white light. Just stopped and hovered there in sky till the jet had passed, moved then, zig-zagged a few times and became stationary again after half hour. I got bored watching in and went in.

Also witnessed an airplane-type drone on July 27, right before dark. Had forgot to take trash down. At the end of my drive, I got out of truck, grabbed the trash from the back, and when I sat the can down I noticed that almost directly above me was this drone. It made no sound, and there was one red light under each wing, and underneath the center of this craft was a round blue dome, lit up. I seriously thought something was taking a pic of me. I stood looking at it with amazement thinking "what next?" I don't think this was a coincidence. It felt timed out to me. it didn't hover, it just kept going at what I'd call a slow speed in the direction of my old camera spot. Odd don't you think? I guess the height in the air to be only a few hundred feet. The wings were about three to four feet long each, and the belly just over a foot. It was silver metallic in color. I heard nothing. I think these craft are on the move again down here."

November 1, 2015

"At 5:50pm I was lying in bed and began to hear strange rumbling coming out of a south-southeast direction, and it came right over my house and lasted a few minutes. I didn't get up to check it out, I was just kind of mesmerized as what it could be. The rumbling was pulsating in my ears and they were slightly painful for about a half hour after going over the house. After the rumbling stopped, I began to hear it again, but it was not as close this time. I got up, opened the front door and saw two craft of some type. Same rumbling sound. I saw two widely separated blinking lights. One was white and one was red. When the white light stopped blinking the red one blinked. The first craft was followed by another craft to the right of it very close, only a blinking red light could be seen. Then I noticed another craft in the northeast closing in out of the north-

The trail camera picked up multiple balls of light and perhaps a small being (circled by Carly)

Enlargement:

northeast. It was going towards the first two craft. This craft also had the white and red blinking lights. The crafts seemed to be very large and could be seen with just the eyes for some distance. The sound they made was not the sound of jets, but a weird pulsating rumble! My ears this morning still hurt from the sound. Did anyone else report this strange incident? Note that the red lights were very large."

November 5, 2015

"I found out that my neighbor did hear the same thing. Same night. She, like me, didn't know what to make of it. She didn't notice the pulsating sound. She didn't get up ither to check it out, which I cannot figure out. I just lay there stupefied. I figured it was hovering above my house maybe for several minutes. and when I did get up I heard it was out front. I believe it was the same craft. Judging just from the distance between the white lights, I'd say as wide as the old B2 bombers. very slow moving, slow strobing white and red lights. I've checked everywhere I could but found any reports. My left ear still hurt in the morning. felt like there was something in it. The pain went away by noon."

July 7, 2016

Carly was out walking in the woods when she ran across a large footprint in the soft ground. Since her property is marked there should have been no one on her property without permission. She believes that this is a print left by a bigfoot.

July 9, 2016

"I picked up a hitchhiker today. This is something I never do, so can't explain why I did such a thing. He was a twenty something year old male, and the whole ordeal was quite strange. His name was TJ. When I went to shake his hand, I wanted to jerk away immediately! Visually it was a hand, spiritually it was NOT!! He didn't know what I experienced in touching him. Spirit was in control of the situation. we showed no shock or fear. Spirit wanted me to experience this. Weeks later Linda Howe had a show on YouTube that explained what I went through, another lady felt what I did in touching a young college girl. She said the college girl was a hybrid. There's more!"

May 28, 2018

This email is out of order but is related to the bigfoot print:

" I was told today that a woman took a boy fishing about 15 miles north from Piedmont yesterday, at a place known as stillhouse hole. While they were there, they smelled an awful odor, and then something across the river started throwing rocks at them. I'm working on getting more info. The woman works at a little restaurant in Desarc Missouri, which is a 15- minute drive from Piedmont. Needless to say, they quit fishing and got out of there. My cousin knows the woman and is going to see if she will talk with you. Are you interested? Bet there are tracks over there. Sand and mud along the river!"

August 22, 2017

Carly contacted me to say that she was outside and suddenly had the feeling she should take photos directly into the sun. This is something she never thought of doing before. She captured a black object in front of the Sun in the photo on the next page.

One day, Carly called me. She was extremely frightened by something that occurred the day prior. Her son said that a man they knew as a business owner

An oddly shaped black object in front of the sun. Photo by Carly
Note, the round object below is likely a lens flare.

in town knocked on his front door. The man had never visited before, and her son thought this was very odd, but asked him to come in and have a seat. The man, who shall remain nameless to protect his identity, said he had something important to tell and to listen very carefully. The man said that he used to work for the government and that he "knew what these people were capable of, and that they could to horrible things to people who saw things they shouldn't."

Carly's son sat there in disbelief because he never spoke of the things that were going on at their property with anyone other than his mother. His mother, in his mind, was the primary witness to all of these strange accounts.

Then the man started shaking and crying. He told Carly's son to be very, very careful and to keep a watchful eye out, and if at all possible that his mother and he should move far away. He did not ask if they had seen or heard anything unusual, or talk about UFOs or aliens, or even sasquatch. All

Close-up of object in front of the sun

he did was give a firm warning, then got up and left, leaving Carly's son dumbfounded. He immediately walked to his mother's house and told her what happened. That evening, and for two weeks after, he and Carly sat outside keeping watch. That is when they noticed that between midnight and 3:00 am each morning they saw what looked like flashlights in the woods. Finally, Carly got a pair of binoculars out and the two of them sat on lawn chairs in a dark area and kept watch. Sure enough, the flashlights returned, and through the binoculars and under the full moon they could see men creeping around in the woods and they were dressed all in black with black face coverings. This disturbed Carly and she called me to relate these strange events, saying "This is private property and there should be no one on it except for us." She questioned who these people were, and after hearing what the local man told her son, they believed that it was likely a military operation. What they were looking for was the big question.

I asked Carly if she could possibly move away and she said that she did not have the means to do so, and besides, this land had been in their family for generations and she was planning on leaving it to her son. She also mentioned that someone from the county government had offered to purchase her property several times, and she refused each time, but she never told me exactly who made the offer. Whoever it was never gave a reason as to why they wanted the property.

After viewing the people in the woods for a while, they did not appear one evening, so Carly's son headed to his house. He had a bad feeling and could not sleep, so he got up and went outside, several yards from his house, and hunkered down in a low-lying area and hidden by brush. Then he heard foot-

Trail camera photo captured 5-27-2012 showing an unexplained bright beam of light and balls of light from unknown sources. Something triggered the camera. Below: enlarged area

Possible bigfoot print

Possible bigfoot print with ruler

Photo: Carly

Possible manipulation of trees by Sasquatch or ETs:

A tree bent over and twisted in an odd fashion. This is something that is typically believed by researchers to be done by Sasquatch. Photo: Carly

Another tree configuration, possibly by Sasquatch.
Photo: Carly

steps coming towards his house from the woods, and suspected that it was the same people they had been observing nightly. Sure enough, he watched six or seven men dressed in military black ops with automatic rifles in their hands, head silently towards his house. One man passed inches from his face without detecting him. They had night vision goggles on and did not use regular flashlights, which would have given away their positions. Carly's son held his breath as the men walked by. They went to his house, looked into the windows, then met at the back. One man said "Damn, he is not here!" Carly's son knew at that moment that the men were there to do him harm and perhaps kill him. After the men left, he headed to his mother's house at 5:30 a.m. and woke her up to relate what occurred in the early morning hours.

Carly called me later that morning, and told me what had occurred overnight and that she was now afraid for their very lives, although neither of them could understand why. I kept the bulk of her reports and photos private, except for one poor quality video of either a man in a monkey suit or a sasquatch walking some distance away from her trail camera video. I had recently asked her to turn the camera on video instead of still photos. It is still on YouTube.

Tragically, this story ends badly. A few weeks after the aforementioned incident, Carly called me. She was terribly distraught. She said that just three days prior to her call she was standing at her kitchen window in the morning when she saw her son's truck driving furiously towards her cabin. She immediately knew that something was terribly wrong, and instinctively called the police for help. She watched as her son turned in front of her house and headed straight for the largest tree on the property, which had large limbs. Somehow, Carly knew that the unthinkable was about to happen—and it did. Her son drove the car under a limb, then got out and hopped in the bed of the pickup. He then picked up a rope, and made a noose out of it, then tossed it over the limb. He tied one end to a part of the truck, put the noose over his neck, and jumped off the side of the truck just as law enforcement arrived. They immediately ran over to Carly's son and held him up while loosening the noose, but it was too late. He was already gone.

Our conversations and emails slowed down after that. Carly would occasionally call to tell me about strange events, but always talked about how she thought that somehow, the military was responsible for her son's death. She speculated that there was some type of mind control involved, although she had no clue as to how that may have occurred. She said that her son was not suicidal, was happy with his life, and enjoyed the peace and solitude that their property had to offer. She also told me that she was going to leave her property to a researcher so they could investigate the property, but later changed her mind because she didn't want anything bad to happen to them.

Carly then began telling me that her longtime partner, (I'll call her Liz), was acting very strange. She began to accuse Carly of moving objects around the house, stealing things, and worse—that she was the cause of her own son's death. At times, it seemed to Carly that Liz was possessed. Liz then left Carly and moved out, only to return several weeks later outside with a large knife in her hands and yelling that she was going to kill Carly. Carly had changed the locks, so Liz could not enter the house, and she eventually left. Carly asked me if I thought someone

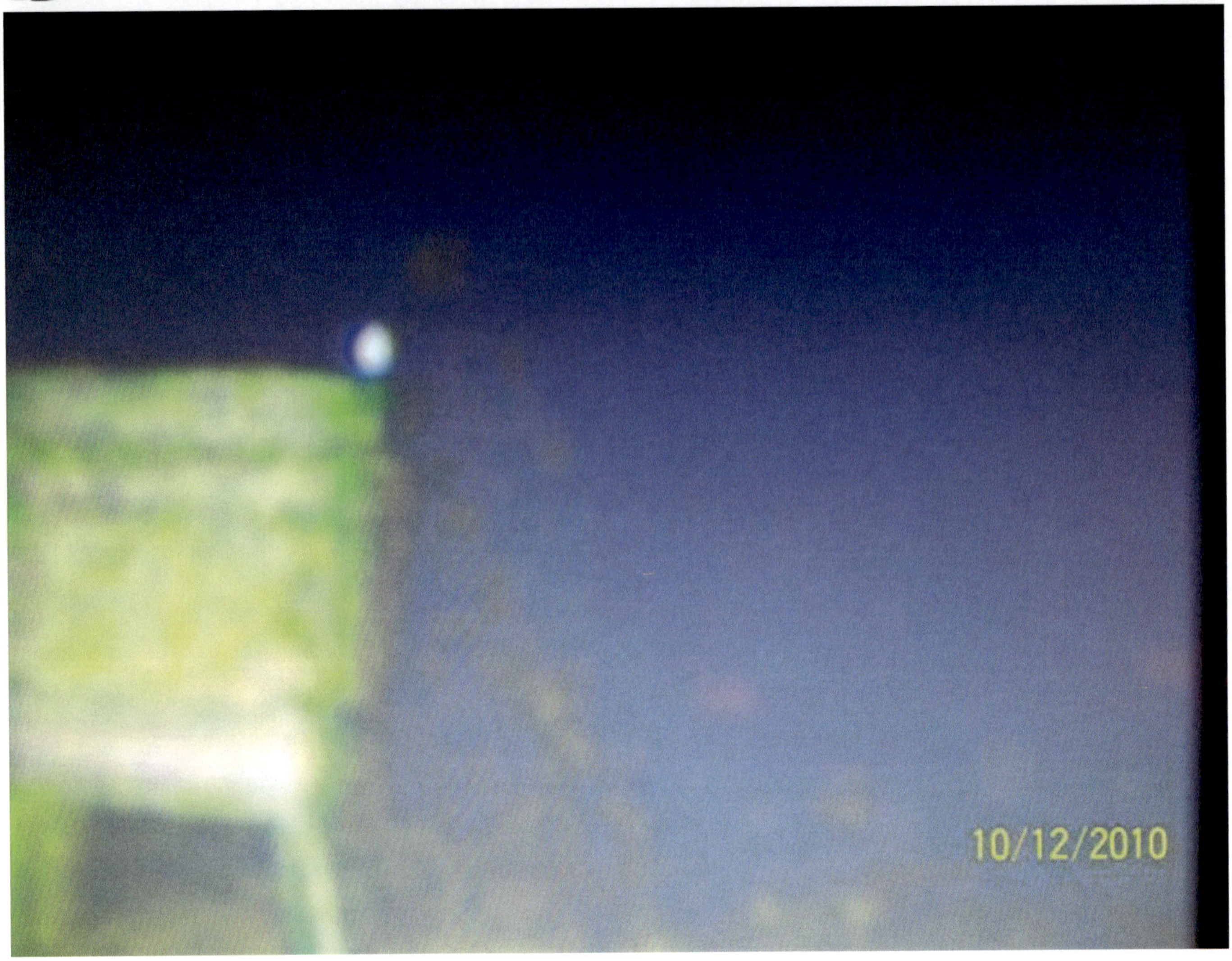

The trail camera captured an orb over the bin and a possible entity standing behind it.

had mind control over Liz. I said I did not know, but that it was probably time for her to move away since the situation had become so dire.

Then COVID hit in 2020 and everyone pulled back. Carly and I never spoke again. I tried to call her but her phone line was disconnected. I finally found out that she died in 2021, however, there is no obituary and I can find no record of her death anywhere. I also cannot find out who owns the property. It would be interesting, and not surprising, if it were a Government agency.

What is the conclusion to this sad story? Carly felt that there were other-worldly creatures on her property related to the multiple UFOs she saw, and that some arm of the government did not want this information to get out. She felt threatened and unsafe. Unfortunately, there are more questions than answers. Why was there such strange activity on this property? How can the UFOs, extraterrestrials, beams, orbs of light, strange animals, bigfoot and military presence, and death of her son (and perhaps hers) be explained? Perhaps we'll never know. I do know this—I'll miss my conversations with Carly. I believe she had something very strange going on at her property.

Carley sent these photos to me in 2010 after she captured images of strange objects Note how a round orb or light comes over the mountain and next to an unexplained light rod

Photo 1

Photo 2

Photo 3

Photo 4

Photo 5

Photo 6

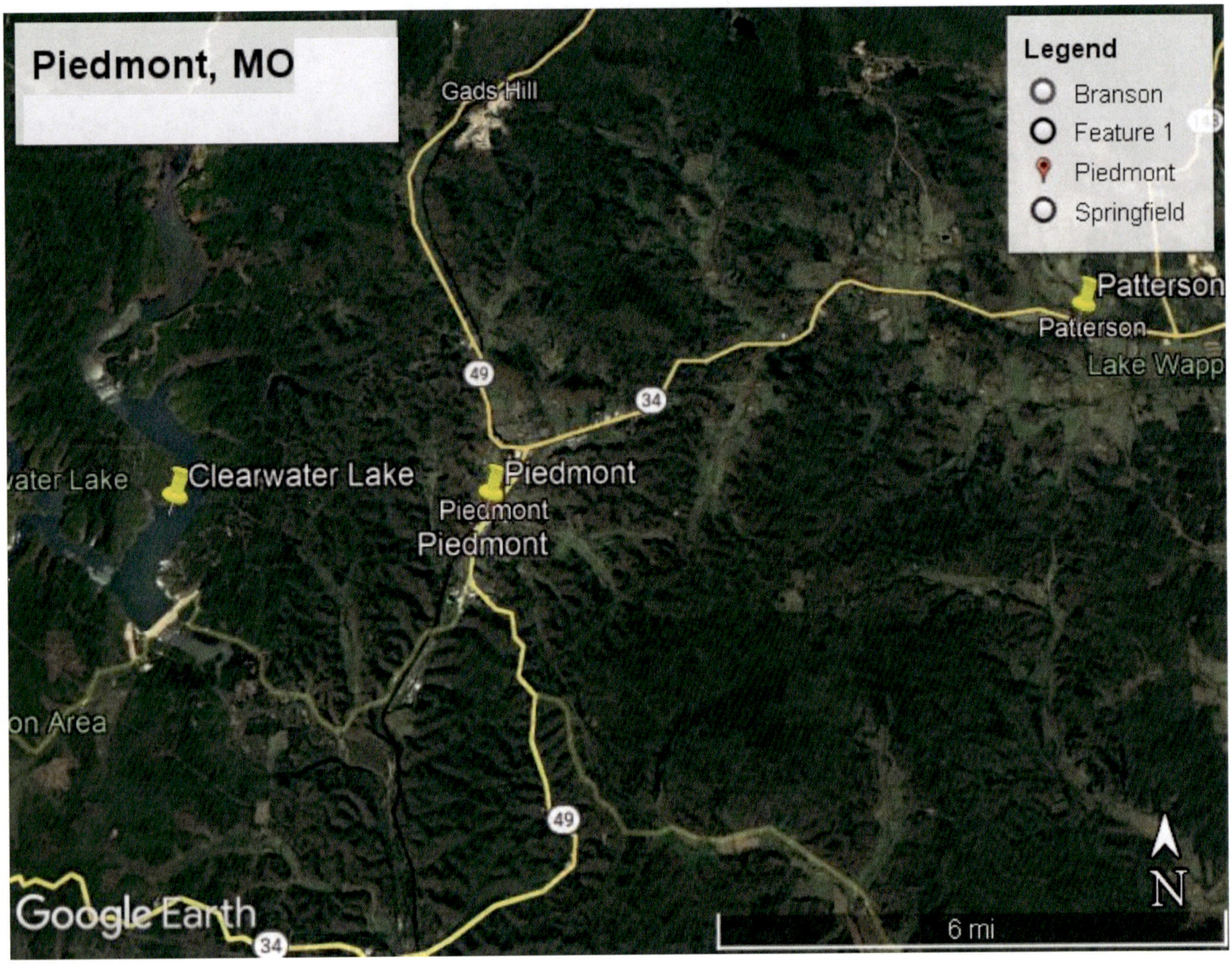

Locations of Patterson, Piedmont, and Clearwater Lake.

Photo: Google Earth

Chapter 9

Two Investigators Have Encounters of their Own

By Margie Kay

In 2011, I was asked to speak at the UFO Revival in Patterson, Missouri as the Keynote Speaker. My talk was about a very strange case in Butler, Missouri, which involved abductions, missing persons, paranormal events, and the military. It was a very sensitive subject and one that I had not presented publicly prior to this event. There was even another investigator from a different organization who involved herself in this case and gained the trust of the witnesses, then told them tall tales about me so she could steal the case, which she did. It also involved several other investigators who, along with me, witnessed some extremely bizarre things that we still cannot explain.

My husband was unable to go with me so I asked a new investigator if he would like to go along. I had a feeling that I shouldn't go alone on this particular trip, even though I have made many travels by myself for years. We were to meet Debbie Ziegelmeyer at the festival site and set up a booth for MUFON together, then both Debbie and I would speak at the festival on Sunday. Our adventure would prove to be anything but mundane, and we had no idea what lay in store for us prior to our departure.

From my report:

"I am the Assistant State Director for Missouri MUFON, and a Star Team Investigator. The person who went with me was a fairly new investigator for MUFON. We had a spectacular UFO sighting, missing time, and possible abduction while on the last leg of our trip from Kansas City to Piedmont, MO, located in the SE corner of Missouri in Wayne County, and again on the way home.

The duration of the trip to Piedmont, Missouri

should have been six hours according to my GPS. With stops taken into consideration, we should have arrived at 8:30 p.m., but we arrived at our destination at 10:37 p.m., so we had over an hour of missing time. We cannot account for the missing time. I had mapped out a route using Map Quest and used GPS, however, the GPS quit working and my passenger, who was a tactical driver for the Army, took over navigation by using my paper map of Missouri. Somehow he got confused and ended up directing me to drive far to the east side of the state. I should note that I have done a lot of driving and am very good at telling what direction I'm going, but in this case, I lost all sense of direction and thought we were driving south instead of east. I will call the other investigator "Tom" since he wishes to remain anonymous.

At about 6:30 pm we decided to stop for dinner near Farmington, Missouri and found the one and only café for miles. It was odd - the café looked like it was a relic from the 1950's, and everyone literally stopped talking and turned to stare at us when we walked in. It was like being in a movie back in time. Our clothes were out of place. I felt uneasy and could not eat much, so we left quickly.

As I was driving South on 67 Highway past Farmington at the very South edge of Francois County, I suddenly had an uneasy feeling that there was something in the distance to my left. I looked in that direction and saw three bright lights in a triangular pattern in the to the East (left side of the vehicle) and across the Mississippi River. It was not a configuration of stars that I recognized. I realized that it was likely a craft rather than three separate lights, and mentioned this to Tom. The object was approximately the size of a penny held at arm's length. The object was much brighter than any planet (5 x brighter than Venus) and was bluish/white in color. It was at approximately 50 degrees off the horizon and east northeast of our position. It appeared to get larger and clearer as I watched it. The uneasy feeling, which started just before I saw the object worsened. Tom then plugged in a device with someone talking and I asked why he did that. He said it was to calm me down. I said to turn it off as it was just annoying me further.

I asked the other investigator to take a look at the object I'd noticed, so he opened the sunroof of my SUV and watched it for some time. He said the object moved to the north behind us, and when I glanced to see it, the object was gone from the east.

He watched several shooting objects without trails moving from West to Northeast, but not behaving like shooting stars. He said, "Oh No," and I asked "What?" and he said, "Those are not shooting stars." Then he said three long white lights appeared, moving very fast from a NE to SW direction. I panicked a bit and began to drive faster on the highway, which was very dark and had absolutely no other vehicles on it for many miles. This concerned me as I knew that this sort of thing occurred to persons who were about to be abducted.

The first triangular object appeared and disappeared three times. The two objects moved around the sky for approximately fifteen minutes and were apparently shooting something at each other that looked like comets or missiles streaking through the sky, according to Tom. From my position I could not see the shooting objects. The skies clouded over, and he lost sight of the objects. We were both very uneasy about the entire sighting, and thought it was coincidental that this occurred while we were on our way to the Piedmont UFO Festival.

We arrived at our destination an hour and a half later than we should have, so considering the strange stop at the café, there was over an hour of missing time that could not be explained. When we finally arrived at the home of the woman who was sponsoring the festival, we entered the house and I asked to use her restroom which was located at the end of the hall. While in the bathroom I saw someone looking in the window in the back. I then asked our host if she had any neighbors who might be looking in the window, but she said that was impossible since the window was two stories off the ground in the back of the house. I realized at that moment that I had just seen a grey extraterrestrial rather than a person, and looked back but it was gone.

The host showed us to our rooms and mentioned that another speaker was sleeping in the loft above us. I immediately fell asleep, which was odd given the circumstances. I was very concerned about our missing time, our getting lost, and the ET in the window.

The next day we went to the festival at the Patterson Community Center and set up a Missouri MUFON booth. Debbie Ziegelmeyer met us at the site. Neither Tom nor I had ever been to this town, although I had long wanted to get there to interview people about the 1970's UFO sightings and to see if anyone continued to see UFO's in the area. We began taking items inside for our booth when Tom asked if I wanted my computer. I said no, because I didn't need it until the next day. He put it on the floorboard of the back seat and put a cloth over the top of it to conceal it, then I locked the SUV and we started to go inside the building to set up. As we were doing so, Tom came inside with some things from my vehicle and said that he had just seen the ghost of a soldier outside and that the spirit seemed to look directly at him, extending his hand for a handshake. This was a surprise, because we were not focused on or thinking about ghosts, and Tom said he had never seen a ghost before in his life. He was a bit excited about it, and I said we should return later that night to investigate the site.

During the lunch break I interviewed several area residents on camera. One man said that his family has been seeing strange balls of blue/white or white light on their property for at least three generations, and that they've never been able to figure out what they are.

A woman told me that she was on the dam at Clearwater Lake when she and her husband saw multiple red lights just under the surface of the water, moving slowly together, and that this occurred only three months before my visit.

Several other people said that they've been seeing unidentified flying objects, some within yards of their position for many years, and they all said that the UFOs never left but that people don't call the police or media any longer for fear of ridicule.
We all went to dinner together with our hosts, and Debbie said she was tired and headed back to her hotel room to get some rest.

Tom and I then headed back to the community center at dusk and parked in the back parking lot, which consisted of a large clear area with two posts and a heavy rope between them to mark the center where cars should park on the grass.

Immediately upon arrival, I grabbed my Nikon camera and walked out to the field behind the community center. I could sense that there were spirits at this location and went into a light trance state so I could see more. I was surprised to see soldiers in blue uniforms hunched down as they walked up a creek on the right side, and a group of men in grey uniforms walking silently down the hill towards us in the back. I heard several first and last names called out, and Tom wrote them down. They were each old -style names. Then both of us saw a lantern swaying back and forth in a field next door. It looked like someone was holding it while walking, then it simply disappeared. Both of us thought that it might have been a spirit. so I took a photo.

Next, to our surprise, the rope between the two posts began to move violently back and forth. Neither of us were near it and no one else was on the site. We aimed our flashlights into the dark and saw no one.

I snapped a few photos of the area with my Nikon camera using a flash but did not review them until I arrived home later. What I caught on film was astonishing (see photos on next pages).

The next morning Tom was speaking to a woman who was a local historian, so he introduced me to her. I told her what had transpired the previous evening and she said "You know that this was the site of a Civil War battle, right?" We both said we had no idea that was the case. Then Tom pulled out the list of names I mentioned the night before, and she said that there was a graveyard right next door which was from that battle, and there were many graves there from both the North and South armies. She then took the list and went to the graveyard, returning to say that all of the names I mentioned were on gravestones at that site! Even I was shocked at the accuracy, and I do psychic/remote viewing work all the time.

But we were in for an even bigger surprise later that afternoon. At noon, Tom went to my car to retrieve my laptop for my presentation later that day. He returned to say that it wasn't there. The vehicle had been locked the entire time, so I couldn't understand why it was gone. But we were able to retrieve the presentation using his laptop and the "Go to my PC" program. In the meantime, the hosts switched my speaking time to 5:00 pm as the Keynote, and Debbie's time to 3:00 pm. This was not announced in advance. The event was going very smoothly, but at exactly 3:00 pm all of the electricity went out.

A man came in to the hall and said that all of the electricity in the entire town was out with no explanation. The day was sunny, and there was no wind or weather event to explain the outage. The event coordinators moved quickly to hook up a generator so Debbie could do her presentation. Then at 5:00 pm, the electricity came back on. We all thought that it was extremely odd regarding the timing of this event, and thought that it might have been related to the presentation I gave about a case I had in Butler, Missouri which was a very sensitive case that involved UFOs, alien abductions, and the military. We thought that someone did not want this information to get out, and the missing laptop and electrical outage were not mere coincidences. I also noted during the presentation that a person I suspected to be

UFO REVIVAL
Commemorating the 1973 UFO Flap
June 17 & 18, 2011
AT THE PATTERSON COMMUNITY CENTER
PATTERSON, MO (5.5 MILES WEST OF HWY 67 OFF HWY 34-BEFORE PIEDMONT)
CREATING A PROGRAM TO INCLUDE THE FOLLOWING EXHIBITS & ACTIVITIES WITH YOUR HELP:
•GUEST SPEAKERS
•COMEDY DRAMATIZATION
•UFO BOOK SALES & memorabilia
•COMMUNITY PHOTO EXHIBITS
•NASA EXHIBIT
•ROBOTICS DEMONSTRATION
•ASTRONOMY CLUB ACTIVITY
•STORYTELLING
•UFO SCULPTURE EXHIBIT BY AREA SCHOOL CHILDREN
•MC'S TO CONDUCT PROGRAM
•SOUVENIRS
PCC STAFF TO SERVE FOOD
40' X 100' GYM WITH STAGE
"CAAW" EVENT
Cultural Arts Alliance, Wayne Ozark Region, Inc.
P.O. Box 643 Piedmont, MO 63957
www.caawcaaw.com
muleicorn@windstream.net
573-856-4005
EARTHPROBE
THE COMPLETE STORY OF THE "PIEDMONT UNEXPLAINED". STATEMENTS TAKEN FROM OVER 200 EYEWITNESSES. COMPLETE WITH ILLUSTRATIONS AND UNEXPLAINED PHOTOS.
BY HAYDEN C. HEWES
P. O. Box 1281, Oklahoma City, Oklahoma 73101

a government agent was in the audience, and shortly after, she managed to get the witnesses from Butler to believe lies she told them so I couldn't continue investigating the case.

That evening our host took us to a haunted hotel that a local couple had up for sale. It was a site where U.S. Marshals stayed while they hunted down Frank and Jesse James, and it had quite a history. I had intended to drive home immediately following our tour, but when we arrived at our host's home it started pouring down sheets of rain. Our host insisted that we stay over one more night and leave in the morning, and I relented, although I sensed that something was going to happen if we stayed, so I resolved not to go to sleep. Once inside, Tom asked me to look inside his room and he showed me his toiletry bag, which he said someone had been rummaging through during the day while we were gone.

After getting into bed, I heard noises outside but assumed that it was the other investigator checking the perimeter, as he did every night. For some reason, I forgot that it was raining cats and dogs outside and he likely would not have gone out in the downpour. I was lying in bed, trying to stay awake when I heard a very loud cracking sound over my head, as if a very thick piece of glass broke. At that point, I knew something was happening, but instead of getting up to turn on the light to investigate I fell asleep!

The next morning, I got up at 7:30 am to find my host and Tom sitting at the kitchen table. We chatted for a minute, then Tom and I took off for Kansas City. After a few miles, Tom asked me to pull the car over and I did so. We both got out of the vehicle and he showed me a horrible looking 1" diameter wound on his leg, which he said he woke up with

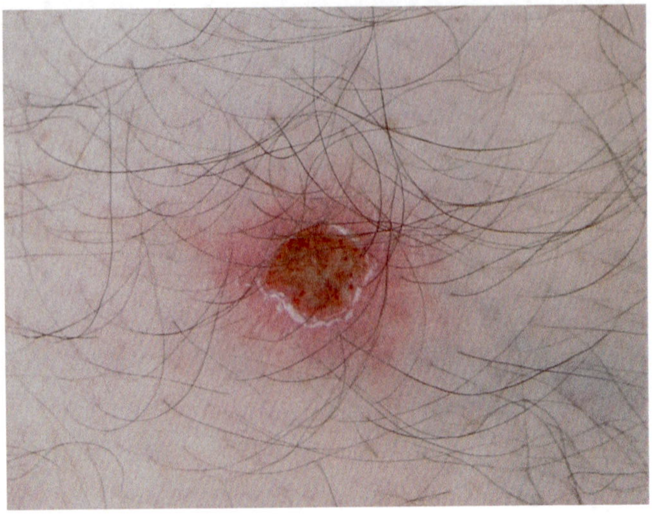

Wound on Tom's leg
Photo: Margie Kay

in the morning and had no explanation for how it got there. At that point, we realized that we had likely abducted during the night. I related the glass-breaking sound, but neither of us could remember anything else.

After I got home, I placed my things on the kitchen table, unloaded my bags and put laundry in the washer. My husband and I went to bed and in the morning I got up and passed the kitchen table, which I cleared off the night before. Sitting in the center of the table was my missing laptop! I checked and my power point presentation was missing off of the laptop.

After this experience Tom quit communicating with me for about a year. I think he was having a difficult time dealing with the events that occurred in Patterson. One evening while my husband was out of town, I heard loud banging on the front door of my house. It was 2:30 am so I was afraid that something had happened to my husband and the police were there to give me bad news. I saw flashing lights outside from police cars. I opened the door and there were police everywhere. One of them had a man and woman and asked me if I knew them. I said no, and they took them away. They had driven a white commercial truck into my driveway, and parked in the back. The police had been following them because there had been a number of stolen commercial vehicles in my town of Independence as of late, and they stopped them as they were gaining entrance to my house via the front door.

I called Tom the next day, and he said that he caught a man and woman trying to break into his front door at the same time on the same night. They both ran and got away. Coincidence? I think not.

At this point, I decided to try hypnosis in order to retrieve any information about the Patterson event , and contacted a licensed therapist who did hypnosis. I made sure she would ask no leading questions. I clearly saw what happened and this is what I recalled under hypnosis:

First incident: While driving to Patterson on the highway south of Farmington my vehicle stopped. Tom and I were sitting in my vehicle and there was a disk-shaped craft hovering above the highway abut 50 feet in front of us. Three tall beings in white robes that covered their faces, hands, and feet walked towards us. I believe that the beings themselves were glowing and the robes were there to cover them so we wouldn't be blinded. I felt no fear, and as one came to my door and opened it I got out and walked with him towards the craft. This being seemed very familiar. On the other hand, two other white-robed beings were having a difficult time with Tom, who resisted going with them.

We walked up a staircase into the craft and the two beings took Tom to the right, and one took me down a hallway to the left. I remember that the walls and ceiling were curved and there was light but I saw no light fixtures. The walls seemed to be made of something resembling a dull stainless steel, and the hallway curved around the edges of the craft. We entered a room where I was then seated in a comfortable chair. My arms were placed on armrests with some type of instruments in them and a female that looked like a hybrid was sitting at a console in front of me, looking at the readings she was getting from the arm rests. I was awake but in some sort of trance-like state.

Next, the male entity who got me out of my vehicle pulled a screen down in front of me from some type of apparatus attached to the back of the chair that went overhead. It looked like a 13" computer screen. On the screen were scenes of things that I assumed would happen in the future, among them multiple volcanoes and earthquakes along the Ring of Fire around the Pacific Ocean, tsunamis around the world on every beach along all coastlines, and a large earthquake hitting the Midwest region. There was more, but I could not recall everything. At no time did I feel any fear and got a sense that these were benevolent extraterrestrials/ Interdimensionals. I recall nothing after that, but obviously they did walk us back to the vehicle, left, and started the SUV again. That explained the missing time on the way to Patterson.

The hypnotist then asked me to look at the last night we spent in Patterson. This time, it wasn't so pleasant.

After the *glass-breaking sound, I was drawn up into an opening in the ceiling and onto a craft hovering just feet above the house. At this point, I could see the craft from outside the house and at least 50 feet away so I saw most of the craft, which was huge. The craft was a medium-gray color, approximately 150′ in width, circular, with structure underneath and protruding squares on the underside perimeter. There were lights around the perimeter and a blue-white light in the center of the bottom of the craft which projected a beam. I assumed that this is what caused the opening in the roof and ceiling.

Next, I found myself lying on a steel table, with straps on my forehead, torso/arms, and ankles. There were three short gray aliens coming towards me with a huge needle and other instruments. I looked towards my feet and there, standing next to a doorway, was a tall brown being that looked half-human and half reptilian. He was obviously in charge and was observing the situation.

Then I heard Tom scream from down the hall and I immediately remote-viewed that room and saw not only several extraterrestrials resembling reptilians, but two human men in military uniforms observing. I knew then that we were in trouble.
I went back to my body down the hall, and as I did so, I noticed the ship control center across the hall. I then stared at the tall reptilian hybrid and thinking fast, sent this telepathic message: "Let us go now or I will use my astral body to go into your control room which is located down the hall and to the right, take over the controls and make this ship crash." I should mention that I have practiced astral travel for over 40 years and have become very adept at it, and likely could have accomplished what I said.

The being looked at me, read my mind to see if I could do it, decided I could in fact do what I said, and he said, "Let them go."
After that, the scene is blank and I don't remember how we got back to our rooms.

As a seasoned investigator, I recognize that what happened to us is a very typical case when it comes to contact with extraterrestrials, and our story is not unlike thousands, or even millions of others, and certainly not unlike what has been happening in the Wayne County area for generations. As investigators we look for evidence that can be duplicated, and in the case of Dr. Harley Rutledge's and our personal experiences, such duplication occurred, leaving no doubt that something very strange is happening in the area.

* The sound of breaking glass has been reported by many who have had encounters with UFOs.

Photos from the Trip

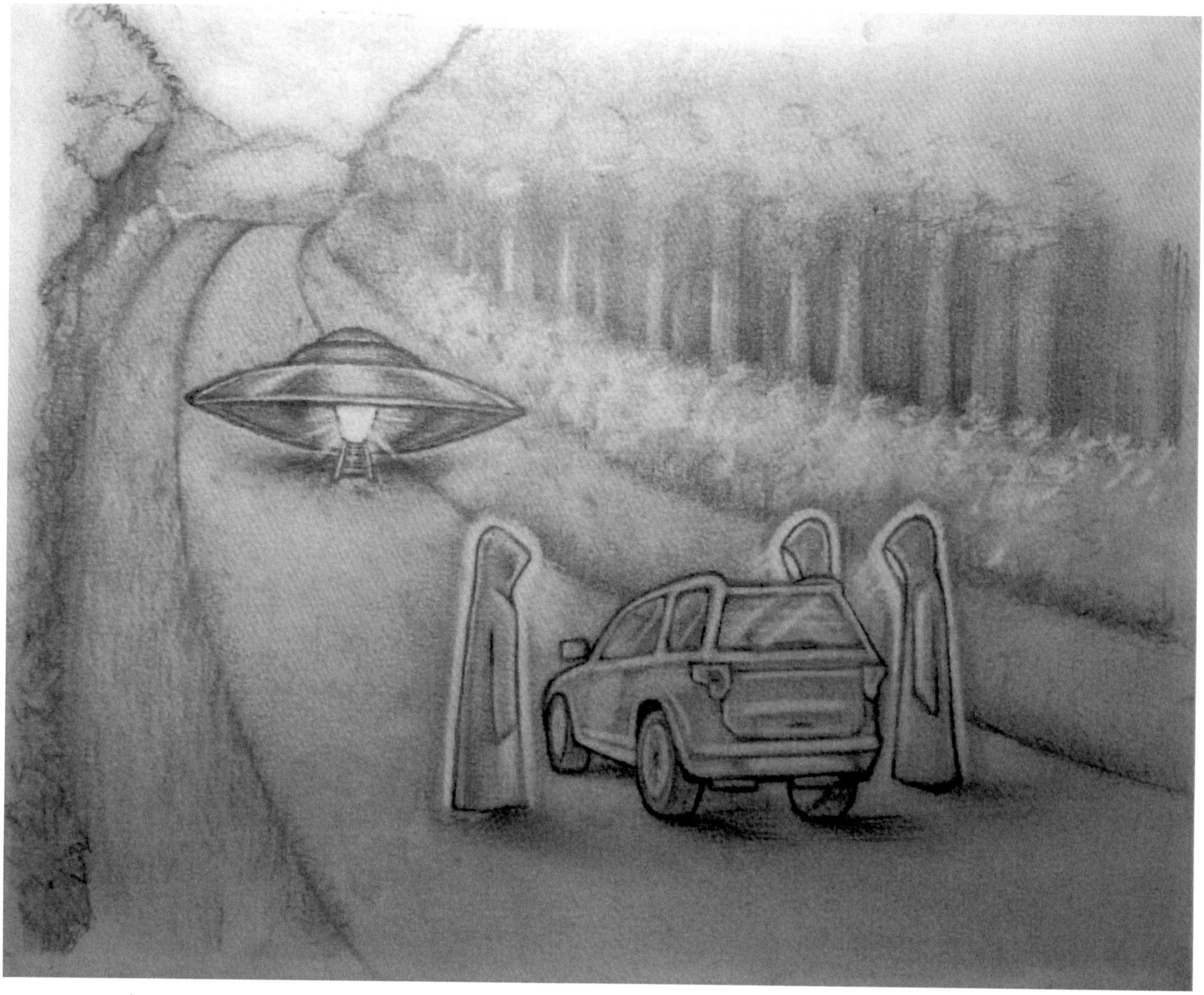

Drawing of scene on the highway on the way to Piedmont, as recalled by me during hypnosis.
Image credit: Kristina McPheeters

A ghostly man with a possible child to his left and unexplained beams of light in back of the Patterson Community Center. These things were not visible with the naked eye. Photos: Margie Kay

A ghost standing in the field to the east of the Patterson Community Center where we saw the lantern moving back and forth. Three other lights in the background were not visible to the naked eye. This was after dark. Photo: Margie Kay

A strange beam of light appeared in the distance, over the city dump. On Google Earth the dump appears to be in a perfect pyramid shape. Photo: Margie Kay

Close-up of unexplained beam of light

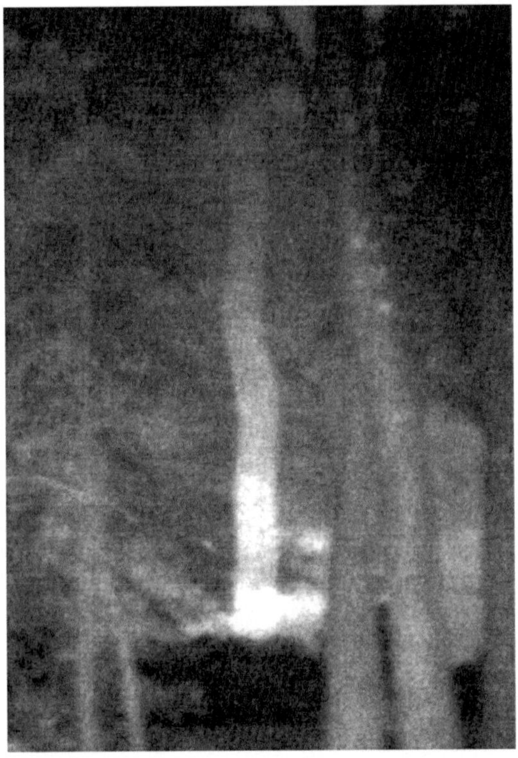

Orbs?

This is the line that was moving violently.

A moth? The shadow should be on the other side, though. The only light sources was the flash from my camera. Was there an unseen light source?

Note: THIS IS A VERY SIMILAR SHAPE AS THE PHOTO TAKEN BY GOOGLE ON THE PREVIOUS PAGE!

Chapter 10

More Interviews

By Margie Kay

Over the past eleven years I've collected a number of accounts from Wayne County residents in person and by phone. In 2011 I interviewed six persons on film while on site for the festival, and in 2022 I interviewed a man in person at the lookout area outside Piedmont. All of the following persons want to remain anonymous so I will use different names for each. These are their statements:

#1. Interview on film with Frank: "I've lived outside Piedmont on a farm my entire life. My entire family, including my parents and grandparents have occasionally seen strange lights in the sky that we could not explain. They are always completely silent, and have no wings or tail, and no blinking red or green lights. The lights are usually white or bluish/white. These are not airplanes or helicopters, so what are they? We'd like to know. I know I'm not crazy, but I can't talk to my friends and neighbors about it."

#2. Interview on film with Carol: "I can tell you that UFOs have been in this area for a very long time going further back than the 1970's, but that was a very busy time. For some reason, the UFOs decided to make a big show of it for all of us to see. Let me tell you, we all saw them, we just didn't all talk about it. But a lot of people wanted to know what the heck these things were so they went public with it. But after a while most of the media wanted to make fun of anyone who said they saw a UFO, so we started to keep our mouths shut after that. It is still going on here, we just don't tell anyone. Things have not changed. The UFOs are here to stay. I don't know what they want, but it looks like they are observing us."

#3: Interview on film with Tom: "I was around in 1973 when everyone was talking about seeing UFOs. I saw two myself. We live just outside town and while I was taking care of a cow on my property I saw some movement in the sky and looked up to see a big round light about the size of a VW floating about 500 feet off the ground. I'd say it was no further than a football field length away from me. It hovered there, silently for about five minutes, then started to move towards me. I backed up a bit, and it seemed to react to me and took off to the south very quickly. I don't think it wanted to do any harm. It was very strange and I'll never forget it. But the UFOs didn't stop coming. A lot of people still see them."

#4. Phone interview with Wendy: I've lived in two different places in the Piedmont area for 46 years and I've heard stories from a lot of people, including my family, so I always keep an eye out or sit outside skywatching on nice evenings. You always see something if you look up. There is something about this place that is attracting them. Maybe minerals in the ground—who knows? But it is something. I've got several videos of blinking and solid silent objects over my property but I don't want to show it to anyone. They mostly look like bright lights, but in one instance I saw something metallic, It looked like a flat disk with no lights on it. The only reason I saw it was because the moon was bright that night. It seems that these things are intelligent, but I don't know what they are."

#5: In–person interview with Butch: Butch took us to the lookout area outside Piedmont. "This is the place people go to hang out and watch for UFOs. Mostly teenagers go out at night when they are looking for something to do." Butch pointed to an area that looked over a valley where UFOs are usually seen. "What they see is lights that act strange. They aren't planes. They don't act like planes. These things are something else, but I don't know what. I've seen them myself a few times. They look like balls of light that are sometimes a circle shape. They move slowly, but can also take off at high speed. I think they are from somewhere else—not from here."

#6. Phone interview with Sam: "My family has lived here for generations and every single one of them has seen something they can't explain. I've seen small balls of light in my yard and big balls of light just above the trees or higher in the sky. They usually come in around dusk—not every night but at least once a week. Sometimes there are more than one of them, too. I've seen up to five at the same time. The lights that come in the yard glow a pale blue or purplish/blue but you don't ever see them lighting anything else up. They are not bright. They'll come in close if you are very quiet, maybe as close as ten feet away from you. I sit outside some nights and think about them and they'll show up. Crazy stuff! That's why I don't talk about it to anyone I know or they'd likely lock me up. I'll bet they see them too as its not likely its just me they show themselves to."

Chapter 11

The Buck Nelson Case

One of the first alien abductees

By Margie Kay

Going back 70 years there is the strange case of a man by the name of Buck Nelson from Mountain View, Missouri, just sixty miles west of Piedmont. Buck Nelson claimed that he was not only visited by space brothers beginning in 1953 but taken aboard their ship several times over the next 12 years. Buck told the story in his book "My Trip to the Moon, Mars, and Venus," which is still available today.

Buck was born in 1895 in Colorado and served in the military. He worked many different jobs after quitting school after 6th grade, including working as a cattle rancher, on the railroad, farming, making lumber and railroad tiles, and in law enforcement. Buck moved around quite a bit for work but when he finally decided to retire, he bought some property near Mountain Home, Missouri where he kept horses and ran a sawmill and farmed before retiring. He lived in a modest four-room house. Buck had a lot of friends and neighbors who knew him well. He was living a leisurely life of retirement when one day his life would be changed forever.

On July 30, 1954, Buck was listening to his radio when it began to go crazy, his dog began barking and his pony was making noise. He headed outside to see what was going on when he saw a huge disk-like object right above his home and two others in the sky further away. He had no idea what it was and ran back inside the house to grab his camera and a flashlight. Buck managed to get three photos of the three objects he saw, but only one photo came out. Buck called the object "Things," because he knew no other name for them and had never heard the term "Flying Saucer." For some reason, Buck flashed his flashlight at the big object, which responded by shooting some type of ray at him and knocked him down. Buck said that at that moment his long-term back pain and neuritis disappeared, and his eyesight greatly improved.

BUCK NELSON

Photo of Buck Nelson from his book "My Trip to the Moon, Mars, and Venus."

Six months later a craft returned and using what Buck believes was a public address system (telepathy?) the visitors asked if he was friendly or not. They asked to land in the back of his pasture where there was a spring, and if they would be allowed to visit sometimes without anyone bothering them. Buck agreed, and the visitors said they would return. At midnight a month later, a spacecraft landed, and three men came to the house and visited for about an hour while the men asked questions about things Buck had around the house.

The visitors told Buck that it was easier for them to come into Earth's atmosphere at midnight due to the location of the Earth and relationship to the sun at that time. The visitors told Buck that if he agreed to share their story they would take him into space. He agreed to go public with this information.

On April 24, 1955 the visitors returned at midnight to take him into space. Buck was allowed to take his dog, Krazy, with him, and leave food and water out for his cat and horse. After entering the craft, he was told he could drive the ship, and although hesitant at first, he did so and had a lot of fun with the controls. The ship could pilot itself, however, so there was no danger of anything going wrong.

Buck said that he was taken to the Moon, Mars, and Venus, along with his dog. He was taken to the ruler's homes on each of the sites and visited with them while they had a meal. Between each stop it was necessary for them to sleep. According to Buck, the entire trip only took a few days, and he did not understand how that was possible. In his book, Buck describes the ship, the people, the planets and Moon and his visits in some detail.

The space brothers also had a dog named Big Bo; it was a huge 350 lb. animal that the space children used to ride. Buck kept hair from the dog and sold it in small envelopes at conventions later. Buck was introduced to a man who was over 200 years old, although he looked like he was in his 20's. Note: Other persons who said they had contact with the Venusian Valiant Thor say he looks like he is 30 years old even though he is immortal. Buck would have no way of knowing that since at the time he was not in contact with other contactees.

Article from the St. Louis Post-Dispatch covering one of Buck's conventions in Mountain Home, Missouri July 6, 1958

Out-of-This-World Ozark Convention

Scene at Buck Nelson's outer space convention. Literature on the subject is on display at right.

BUCK NELSON

Believers Gather at Buck Nelson's Farm to Hear First-Person Stories of Saucer Trips Into Space

By a Staff Correspondent of the Post-Dispatch

MOUNTAIN VIEW, Mo., July 5.

WHEN PEOPLE started seeing flying saucers—or unidentified flying objects—it probably was inevitable that sooner or later they would start banding together, not only to share a common experience but also to form a protective bond against a skeptical world.

Just how many flying saucer believers there are is a matter of conjecture. If you talk to a dyed-in-the-wool believer, he will tell you they now number in the millions and tens of thousands belong to the many space clubs scattered over the country.

They will also tell you that communication between this earth and other planets has progressed much further than you think.

For instance, they insist, there are thousands of beings from other planets living right among us today, passing themselves off as earthlings. You needn't fear them. They mean only good for us. You can tell one, however, when you shake hands with him. Due to the difference in bone structure, his hand probably will feel like an inflated rubber glove. He may vanish, too, right in front of your eyes.

It was inevitable that sooner or later there would grow up within the main body of saucer fans a sort of elite group who have not only seen flying saucers but who have taken rides in them and visited other planets. The elite group is growing every day.

One of the most famous (within space circles) is Buck Nelson, a short, skinny middle-aged farmer who lives alone in a three-room shack in a remote, wooded section of Howell county, near Mountain View, in the southern Missouri Ozarks.

It was just four years ago this month that a space ship landed near Buck's cabin and, after a couple of visits, took Buck to Mars, Venus and the Moon, he said. Since then Buck has become a celebrity among space fans and has made talks in many parts of the country. He has been an honored guest at many space conventions.

Buck's short career was crowned when he held a space convention last weekend at his own ranch. It didn't quite come up to expectations. Buck had expected several thousand people but there were only about 300. In all other respects, however, it was a highly successful convention which measured up to most of them in that there were several persons present besides Buck who had made trips to other planets, they said. One group claimed to have seen a flying saucer Saturday night, the first night of the two-day meeting.

This wasn't surprising to anyone because the delegates made

MAJOR WAYNE AHO

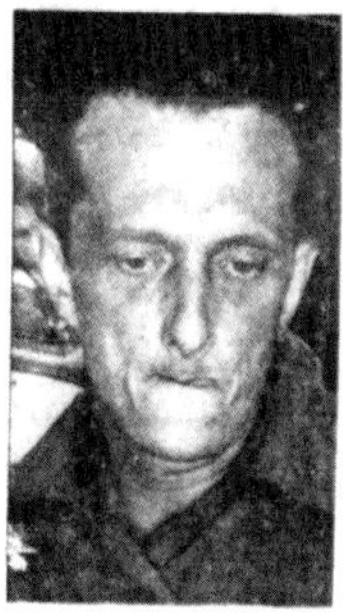

LEE CHILDERS

built out of saplings and scrap lumber, and four booths.

One booth was for space literature. A whole school of writing has sprung up about space and space travel, and on display were no less than 114 different books on the subject. This, it was pointed out, does not include science fiction. They are the works of people who are fighting the battle of the saucers, or first-person accounts by people who claim to have been in them and traveled to other planets. Typical titles: "Men Among the Stars," "Venus Speaks," "Dwellers on Two Planets," "Is Another World Watching?"

Next to the book stalls was a table some 10 feet long on which were displayed dozens of paper bound books, mimeographed pamphlets and other material on the subject.

Next to the speaker's stand was Buck's own concession where one could buy small toys, postcards, ball-point pens, pennants, religious pictures, horns, balloons, pocket combs,

Campers from Affton in St. Louis county pitch their tent on the convention camp grounds. From left are LYNN BENNETT, 15, MRS. J. A LODOLY and her daughter, SHARON, 14.

—By Renyold Ferguson, a Post-Dispatch Photographer.

Convention visitors taking it easy in the shade during a lull in the proceedings.

me with the expenses. I spent $60 getting this place graded, not to mention the advertising."

Buck said he hadn't had a visit from outer space since a year ago Christmas. "They" came down and spent Christmas with him, he said.

The convention had no program, he explained. The chairman just calls on people to tell their experiences. He pulled a big silver watch out of the bib of his overalls. "About time to start now," he said.

dead." The engine stops, the radio stops and the battery goes dead. If this happens you can be pretty sure there are space people about and they are trying to make contact.

A favorite story among the delegates concerns the man who was flying his private plane, and all of a sudden it "went dead." But he didn't fall. He was suspended, with motor stalled, high and motionless above the ground. He looked around and sure enough,

which hovers out in space and which is 2000 miles in diameter. It took the combined best brains of space men from 22 different planets 85 years to build. It's round in shape and can be navigated. Smaller ships land on it by means of a series of air locks.

I tell Childers's story not because it's exceptional but because, at a convention like this, it is so unexceptional. It could almost be called typical of what's happening to a lot of people nowadays. They will

age he receives from the faithful. He is a college graduate and was a major in Army Combat Intelligence in World War II, serving with the First Army all through Europe, including the Battle of the Bulge, he related.

Like a lot of other people, Aho was more or less drawn into the thing out of curiosity. "I got interested back in 1947," he said, "when I started reading so many reports of flying objects. I read all the reports very carefully and I became convinced that a lot of those people especially Army and commercial pilots, were not just having hallucinations. I concluded that the things they were seeing were not of earthly origin."

Until 1957, Aho's interest was academic. Then, while attending a space convention in California, it happened to him. "I saw a craft that at first looked like a star. It came closer and I could see it was egg-shaped. It flashed an orange light. It glided down and hovered about a quarter of a mile away for five minutes. The orange light turned into a brilliant orange-red light. There were others with me watching it.

"After about five minutes it glided down onto the desert and landed. I went out to it, and as a matter of fact I spent the whole night there."

Just what happened and what he saw, Major Aho won't tell.

"People are not ready to accept it yet," he said. "But I did learn that they are just as real as we are. The experience was of such note that I felt compelled to go to Washington and work toward an educational and scientific program and to try to encourage Congress to create some kind of body to gather and really study all the available material."

Aho said he has visited every Representative and Senator. Aho now devotes himself full time to the cause of proving not only the existence of flying saucers but the existence today of inter-planetary relations and communications.

He believes that saucer fans now number in the millions and that there are today thousands of people from other planets living among us.

Well, one wants to know, does he believe that Buck Nelson went to Mars and Venus and that Lee Childers made 21 trips into outer space?

"I won't comment on individual cases," he said, "but I believe many people are giving authentic accounts of their experiences. And I believe there are thousands of others who

While on his journeys, the space people told him that Venus has 17 hours of day and 17 hours of night. They also told him of their laws and rules, and how they have no police, no war, and very little illness because they only eat healthy foods. The space people do not drink alcohol, caffeine, or smoke. But their most important message to humanity was to stop using nuclear energy or bombs or it would mean the destruction of the planet. It is interesting to note that many of the other contactees of the time and later including George Adamski, Dr. Frank Stranges, and others received the same message from Venusian people.

Buck was told by the space people that there were 1,500 Venusians on our planet posing as humans at the time. He said they look the same as us, only more perfect and in perfect health. One wonders if this is true and if so, if there are even more of them among us now.

Buck Nelson gave an interview with a reporter, and afterwards he was contacted by saucer groups all over the U.S. who requested that he speak at their conventions. While in Chicago scientists and astronomers questioned him. One astronomer from the west coast asked Buck what it looked like in space. He replied that it looked Inky-black. The astronomer thanked him and left.

Nelson said he was also visited by people from the Armed Forces, who investigated the site, and even bought the clothes Buck wore while he was on the saucer. They found the buckles and metal pieces magnetized. They even found that the eyelets of his shoes were magnetized. The military investigated the landing site of the craft as well.

Buck was introduced to a human named Bucky, who turned out to be a cousin of Buck's and was taken by the Venusians at age 17 to teach them English. He did this willingly and his parents had been preparing him for this service since he was four years old. Buck saw Bucky on several occasions. During a visit on December 25, 1955, at 1:30 AM Bucky said the following:

"Will tell you why I am here. I have just returned from California, then on to see my folks in Colorado. Now here to see you, Buck, and tell the world on this tape recording that this world must give up ATOMIC WEAPONS AND WARFARE. The next war, if fought, will be on American soil. America will be destroyed, then civilization all over the world will be destroyed. We are here to see which way this world will use Atomic power, for peace or war. We have stood by and seen other planets, one other, destroy itself. Is this world next? We wonder and watch and wait.

Again, I say, give up your Atomic weapons and may Peace be on this Earth.

I will tell Buck much more that he can tell the world. I know that Buck will want my time here to be spent in a private home-like way, and I also desire it that way, so must say goodbye to all the world ...

From BUCKY of VENUS"

Buck Nelson became somewhat of a celebrity, traveling to conventions across the U.S. along with George Van Tassel, George Adamski, Dr. Frank Stranges, Howard Menger, and others. He hosted his own saucer conventions at his property (Buck's Mountain View Ranch) in Missouri for at least 12 years. He captured the attention of Dr. Elmer D. Janzen of Geneseo, Kansas and John Dean of Nickerson Kansas. Dr. Janzen and John Dean attended the saucer conventions and made audio recordings

Photos from Buck Nelson's conventions

Source: St. Louis Post-Dispatch

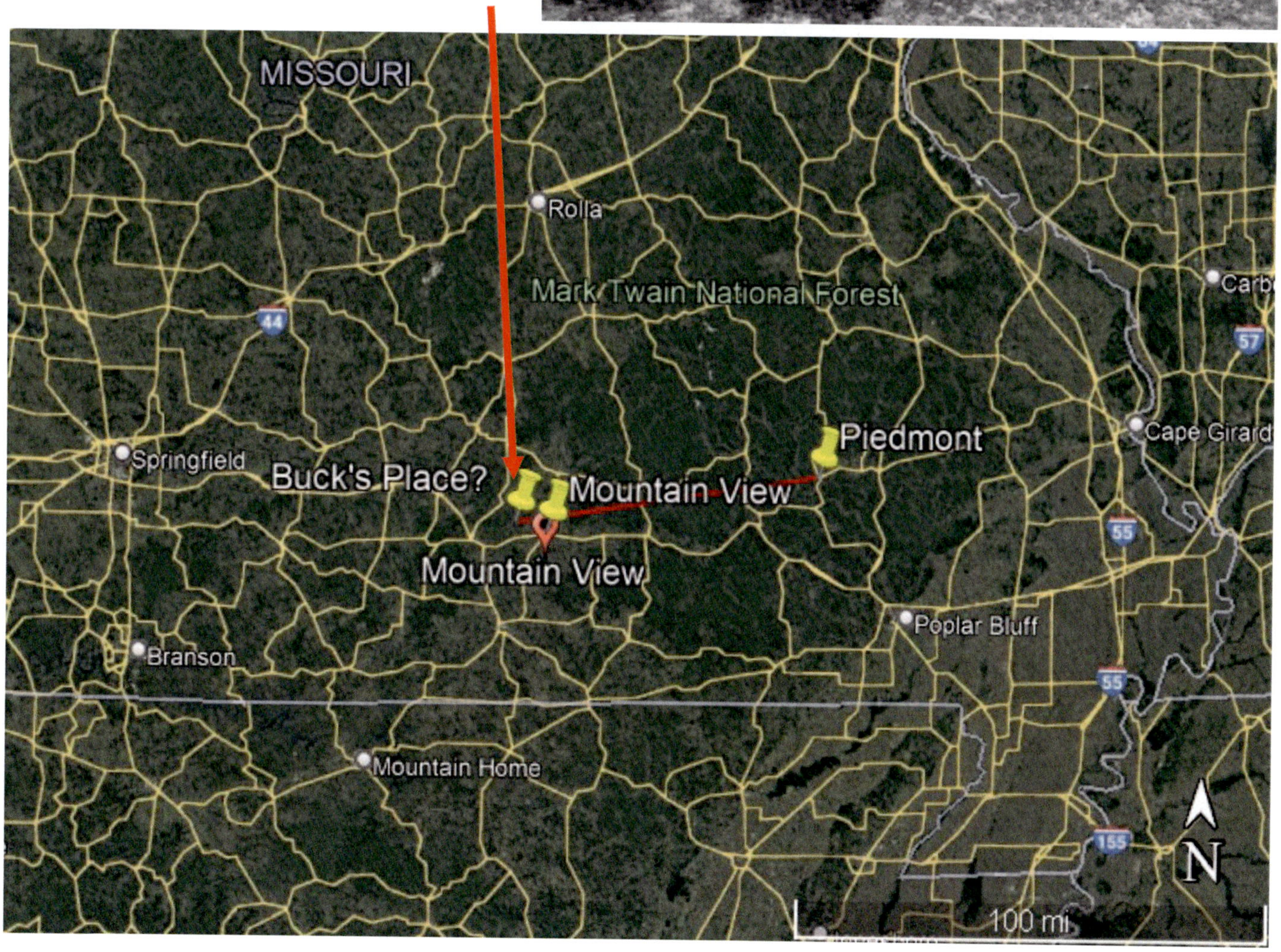

of them, which the City of Geneseo owns. Buck Nelson's talks can be heard in the recordings. I first found out about Buck Nelson while speaking at the first UFO Day festival in Geneseo, Kansas in July of 2022 at the invitation of the president of the Geneseo Museum, Jim Gray. *Note: it is worth a visit to the museum as it has many historic artifacts about UFOs and other subjects.* The city plans to have this event annually in July. I visited Dr. Janzen's Museum and found information about Nelson. The association played an audio recording of one of Buck Nelson's talks, and he did sound genuine.

Photo from one of Buck Nelson's Spacecraft Conventions

When Buck first started hosting his conventions, and invited people to attend at no charge, over 2,000 people attended the events, but in later years less than 200 people showed up. Reporters attended and authored articles. There are multiple articles on the internet starting in the 1950's, but several new articles appeared recently, likely due to the fact that it is the 70th anniversary of Buck's first sighting.

There is no record of Buck's death, and references to it in newspapers are only speculations ranging from 1977 to 1982. There is an article on "Findagrave.com" which asks for information on the death of Buck Nelson. A search at Ancestry.com revealed no information, and an exhaustive internet search yielded no results. Apparently, no one knows where Buck went or when he died. His space friends may know more about this.

In 2019 the property was purchased by Buck Nelson U.F.O. Ranch, LLC.

Thanks to Cowboy Jim Gray, Director of the Geneseo, Kansas Museum for sharing information about Buck Nelson.

Chapter 13

The Great UFO Wave of 1973

By Margie Kay

The year 1973 was notable due to the fact that there were a number of UFO related events that occurred across the United States and in other countries. Besides the Southeast Missouri Ozarks Flap, a number of cases brought the subject of Ufology to the forefront of people's minds due to media attention. Among them:

1. In 1973 famed actor **Jackie Gleason** said that he was taken to a secret facility by his good friend President Richard Nixon. The story broke when Gleason's ex-wife Beverly Gleason wrote an article about the encounter. Nixon had reportedly taken Jackie Gleason through Homestead AFB in Florida to see two deceased aliens with small bald heads. Nixon's diary does confirm that he met with Gleason in Florida. Gleason was so into UFOs that his house was fashioned after one and was called "The Mothership."

2. President Jimmy Carter filed a UFO sighting on September 18, stating that he had a sighting in October of 1969 in Leary, Georgia at about 7:30 p.m. Others with him witnessed the event. They saw a very bright object with changing colors about the size of the moon. After that, Carter vowed never to ridicule people who claimed to have seen a UFO.

4. The "Great UFO Wave of 1973" occurred in Appalachia in middle Tennessee. In the fall of 1973 the "Autumn of Aliens" began in September when a Nashville woman and her three children saw a distant light then a triangular red object with three white lights on it moving rapidly. On September 25 Shelby County deputies in Memphis sighted a hovering craft with two beans of light sweeping the ground. On October 1, three teenage boys saw an egg-shaped craft lad near Anthony Hill in Giles County. The boys claimed to see a large, hairy creature with the craft.

On October 15 a family in Berea, Tennessee saw lights in the woods, and a separate witness saw a being with a glowing white head cross a nearby highway. Newspapers reported sightings in multiple areas until the reports slowed down in November.

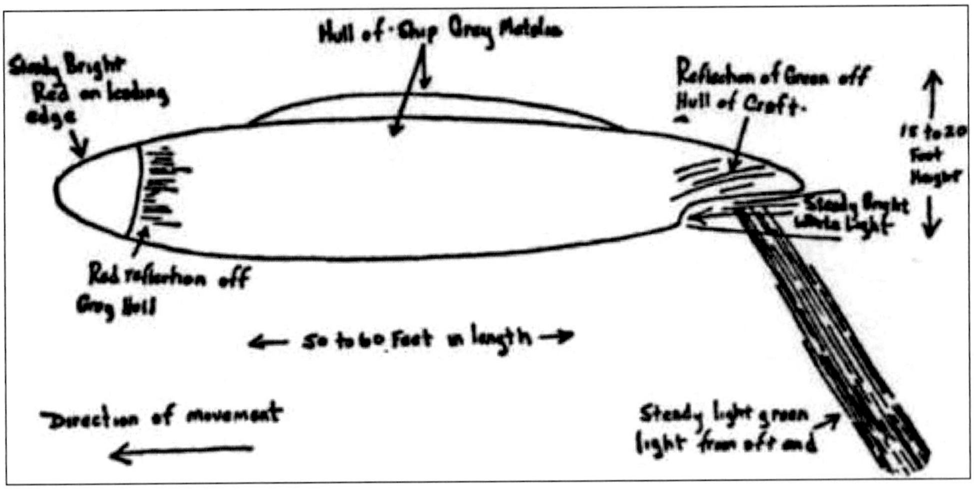

This was a sketch of what Capt. Lawrence J. Coyne and his crew witnessed in the sky over Charles Mill Lake on Oct. 18, 1973.

5. Helicopter dodges UFO: On October 18, 1973 residents on the west side of Charles Mill Lake in Ohio saw strange lights in the sky, but the Army Reserve witnessed unusual lights as well. An Army helicopter was cruising at 1,200 feet when the three occupants noticed a single red light to the west, then a red light to the southeast. As the light rapidly approached the helicopter Capt. Lawrence Coyne descended rapidly but could not outmaneuver the red object. Just as the crew braced for impact the light stopped and hovered above and in front of the helicopter which was now visible as a cigar shaped gray metallic structure with windows. A green beam of light then scanned the helicopter, and seemingly pulled it upwards for about 10 seconds. The crew watched as the unidentified object accelerated to the west, then turned over Lake Erie and moved out of sight rapidly. The compass on the helicopter quit working and had to be replaced.

This case won a prize by the National Enquirer for Best UFO Case of 1973 and it was included in Rod Serling's documentary "UFOs: It Has Begun." The captain of the helicopter was a guest on the dick Cavett Show, and later he made a report to the Special Political Committee of the U.N. He said, " As a result of my experience, I am convinced this object was real and that these types of incidents should require a thorough investigation." FATE Magazine covered this event in August of 1978, and Jennie Zeildman with the Center for UFO Studies published "A Helicopter-UFO Encounter over Ohio in 1979.

6. UFO Mania in Muncie, Indiana: In October of 1973 a county sheriff reported seeing three UFOs according to the Muncie Star. Twenty people called police with similar sightings. The next day, two brothers, 13 and 9, took refuge in a chicken coop when they were buzzed by a UFO in Chester County, Tennessee.

Two days later, according to a story published in The Muncie Star, police found a weather balloon that was likely responsible for UFO sightings north of Indianapolis. For several days in a row, Muncie newspaper readers saw account after account of UFO sightings elsewhere. "Reports rampant in the south," read one headline.

Then, on Wednesday, Oct. 10, 1973, the UFO craze came to Muncie. The evening before, The Muncie Star reported, "mysterious multi-colored lights" were seen by "several hundred" people. About 100 people called police, sheriff's Sgt. Richard Cranor said, over the course of two hours. Another 50 to 60 people called Muncie police. For several days in a row, Muncie newspaper readers saw account after account of UFO sightings elsewhere. "Reports rampant in the south," read one headline.

7. UFO landing in Columbia, Missouri. On June 28, 1973 at 12:30 a.m. James Richard, 41, and his 16-year old daughter, Venea, heard a strange thrashing sound and went to the window of their mobile home to see where the sound was coming from. The two saw to white beams of light , which faded away, revealing an oval shaped object hovering approximately 15 feet above the ground. The area was "lit up bright as day." according to Richard. The nearby trees swayed as if pushed by a strong wind, one limb snapped off. Richard notified police as the lights in his trailer dimmed twice. Investigators found broken tree limes, burned leaves, and imprints two feet deep into the ground. Investigators J. Allen Hynek and Ted Phillips investigated the site. The complete report is available at (missourimufon.org)

8. Was there a UFO crash in Oklahoma? We read an account but have been unable to verify the source of this claim.

9. 1973 UFO Flap in Southern West Virginia. On October 11, 1973 dock workers Charles Hickson and Calvin parker, Jr. reported that they were abducted by aliens while fishing by the river in Pascagoula, Mississippi. The men described gray non-human beings with long arms ending in claws. They were paralyzed and floated onto a hovering craft, before being left on the shore. The two credible witnesses gained national attention after reporting the event to the authorities. Both men passed lie detector tests given by law enforcement. 45 years later Maria Blair n came forward and related that she saw a blue light right over where the men were abducted. Maria met Calvin Parker for the first time at the same spot on the Pascagoula River where the abduction occurred after coming forward with her story. Parker wrote a book about his experience and an historical marker has been placed at the location commemorating the event.

The State Police reported that they received numerous reports of UFOs from all over Raleigh County from 8:30 pm to 11:30 pm. Sightings were reported in seven counties that evening.

Even John J. Gilligan, Governor of Ohio saw a UFO shoot a vertical beam of light. At a press conference he said, "I saw one (UFO) the other night, so help me. I'm absolutely serious. I saw this. It was not a plane. It was not a bird. It didn't wear a cape. And I really don't know what it was."

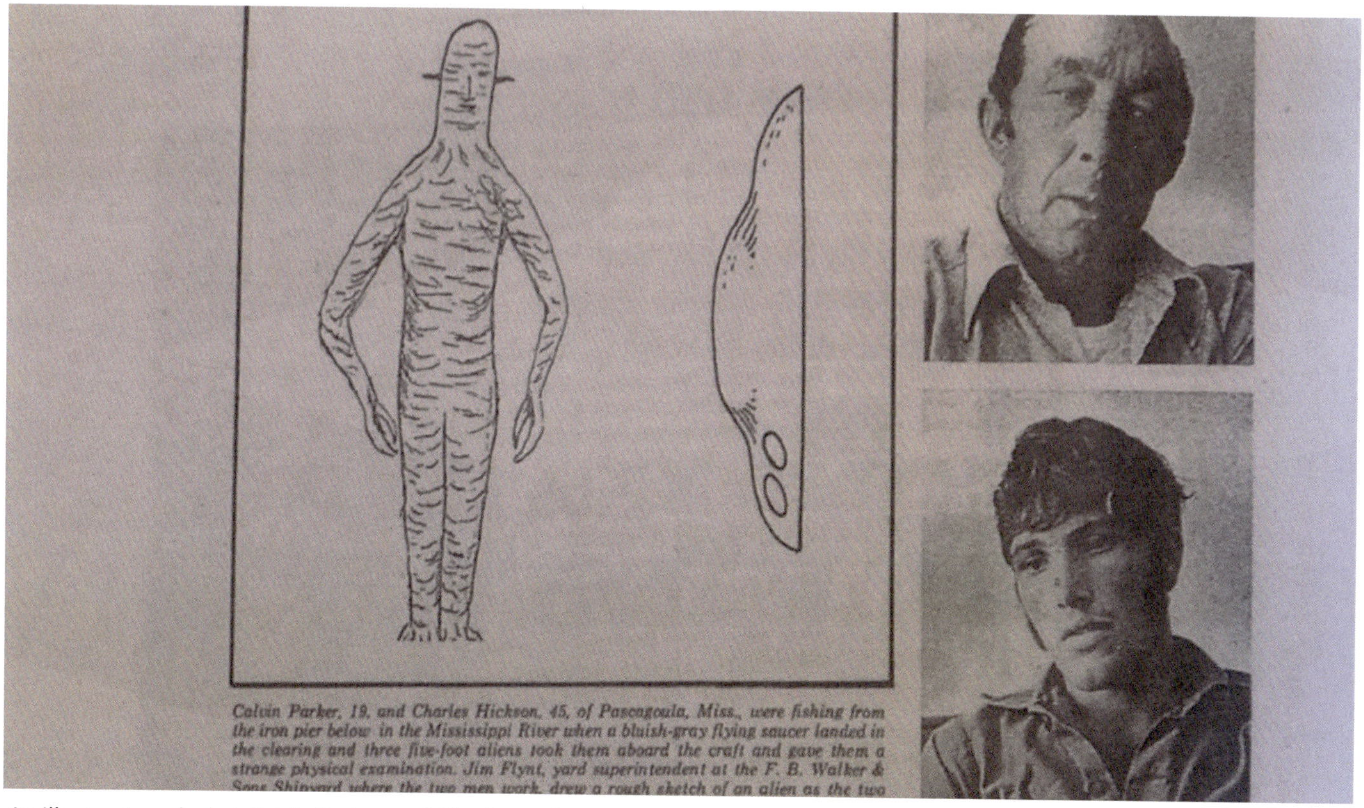

An illustration of what Calvin Parker and Charles Hickson described they saw the night of the alleged alien abduction. Photo: (WLOX)

In his 2018 book, **"The October Scenario: UFO Abductions, Theories About Them and a Prediction of When They Will Return "** Kevin Randle gives a detailed account of many of the 1973 incidents involving thousands of witnesses. People saw UFO's up close or landed. Hundreds sighted occupants riding in craft, and some were abducted.

In New Hampshire, while driving down a highway a man and his son saw a strange object with a tall humanoid standing beside it. In North Carolina a family reported seeing a creature with red glowing eyes, long hair, pointed ears, and a hook nose on a gray face. It could leap 60 feet at a time. A group of teenagers in Savannah, Georgia reported seeing ten black, hairy dogs exit a landed UFO and run through a cemetery.

Randle covers multiple cases in Utah, Argentina, Mississippi, Italy, the Midwest, and more in this book.

Visit www.indiebound.org/book/9780934523356

It seems that southeast Missouri was not the only UFO hot spot in the world that experienced strange unexplained sightings during the year of 1973.

Conclusion

Missouri is known as the "Cave State" with 6,400 caves throughout. There are more than 110,000 miles of running water, and three major rivers flow through the state, which are the Mississippi, Missouri, and Meramec. The New Madrid fault is located beneath the Mississippi River Valley and is one of four major fault lines in the United States. There are also multiple faults throughout the entire state.

Missouri is home of the Southeast Lead Belt and also contains minerals such as quartz, galena, calcite, barite, millerite, dolomite, fluorite, pyrite, marcasite, bituminous coal, amethyst, chalcopyrite, malachite, gold, and barite. Missouri also hosts multiple limestone quarries. Missouri has the location center of the United States population which is Hartville, and is the site of part of the native American "Trail of Tears."

Sightings of unknown craft and unidentified lights both in the sky and moving through wooded areas and across open fields have been documented in Missouri at least since the mid 1800s. The challenging task for UFO investigators is understanding and identifying the alien interest in our area.

Water seems to be an attraction as multiple sightings of USOs in multiple locations. All life needs water to survive—at least on Earth, so it stands to reason that visitors have similar needs. A common theory is that alien craft may be taking Hydrogen can be extracted out of water for fuel, so we must consider that possibility as well. Could the visitors have an interest in water and caves? Recently, divers found a large cave underneath Clearwater Lake that was previously unknown.

One would surmise that the New Madrid fault line would be a deterrent, but many UFO investigators have the belief that UFOs could be getting energy from faults. So-called "Earthquake Lights" are being investigated as a type of energy released by faults.

Lead, crystals, or other minerals could be an attraction because of their abundance, so the more one ponders the possibilities, the more frustrating an answer to the solution becomes. Could it be the combination of them all?

Whatever the reason, sightings of unidentified craft, lights, alien beings and even Bigfoot, have been documented in Missouri for decades. Due to modern technology, sightings are even more detectable than ever before. Reports in Missouri of the unknown are increasing daily and investigators of the unknown with modern technology available to them are able to measure the altitude, distance, size, and speed of many of these objects.

We now have cameras that record in infrared or night vision at high definition, FLIR, EMF detectors, spectrum analyzers and other tools at our disposal.

In conclusion, there is no conclusion. The investigating of unidentified craft and lights in Missouri continues today, and the skills developed by scientists such as Harley Rutledge have been enhanced ten-fold. Today's investigators are discovering answers to unanswered questions of the past and will continue to advance their knowledge through todays available technology.

These are exciting times in Missouri with more to come in the very near future. The answers are out there to be found and todays Missouri Investigators are determined to find them.

Debbie Ziegelmeyer and Margie Kay

In our neighboring State of Arkansas, there is a place called Board Camp Crystal Mine just south of Mena. At this site, multiple UFOs have been witnessed and photographed over the property, and cameras have captured strange levitating rocks on the site. Some investigators believe that the UFOs may be visiting the property in order to extract crystals, which amplify energy.

Why is it that the media no longer covers UFO sightings as they used to in the 1970's? More sightings happen today than ever before, yet the media coverage is slim unless it is released by the military. One could surmise that the media has been told not to cover sightings any longer, perhaps to avoid panic. One thing is certain, the people of Wayne County, Missouri will not be surprised if and when disclosure occurs.

Geological Maps

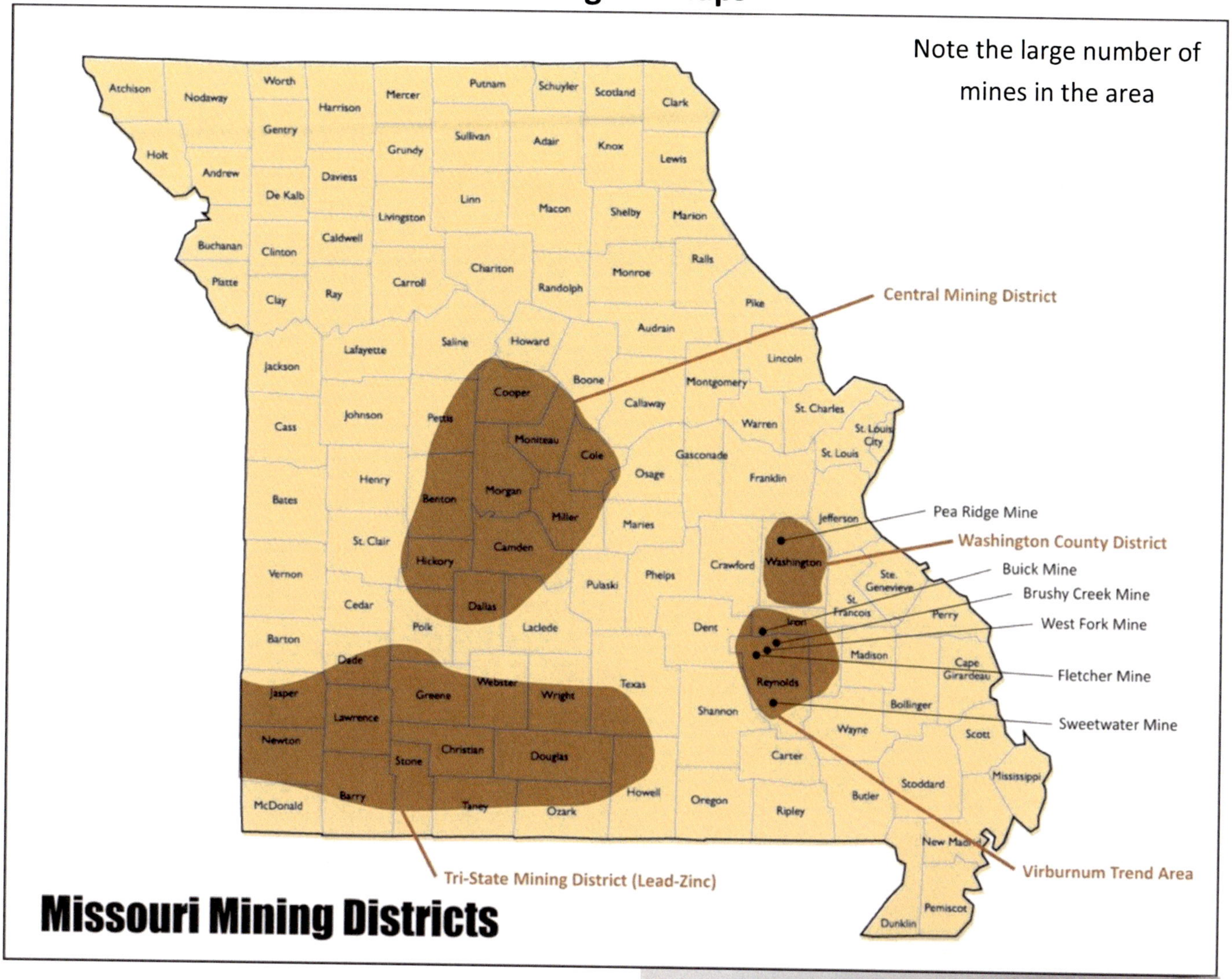

Image: virtual museumofgeology.com

Barite – Lamb Mine, Gravois Mills, Morgan County, Missouri
Image: virtual museumofgeology.com

Galena & Pyrite – Brushy Creek Mine, Oates, Viburnum Trend District, Reynolds County, Missouri
Image: virtual museumofgeology.com

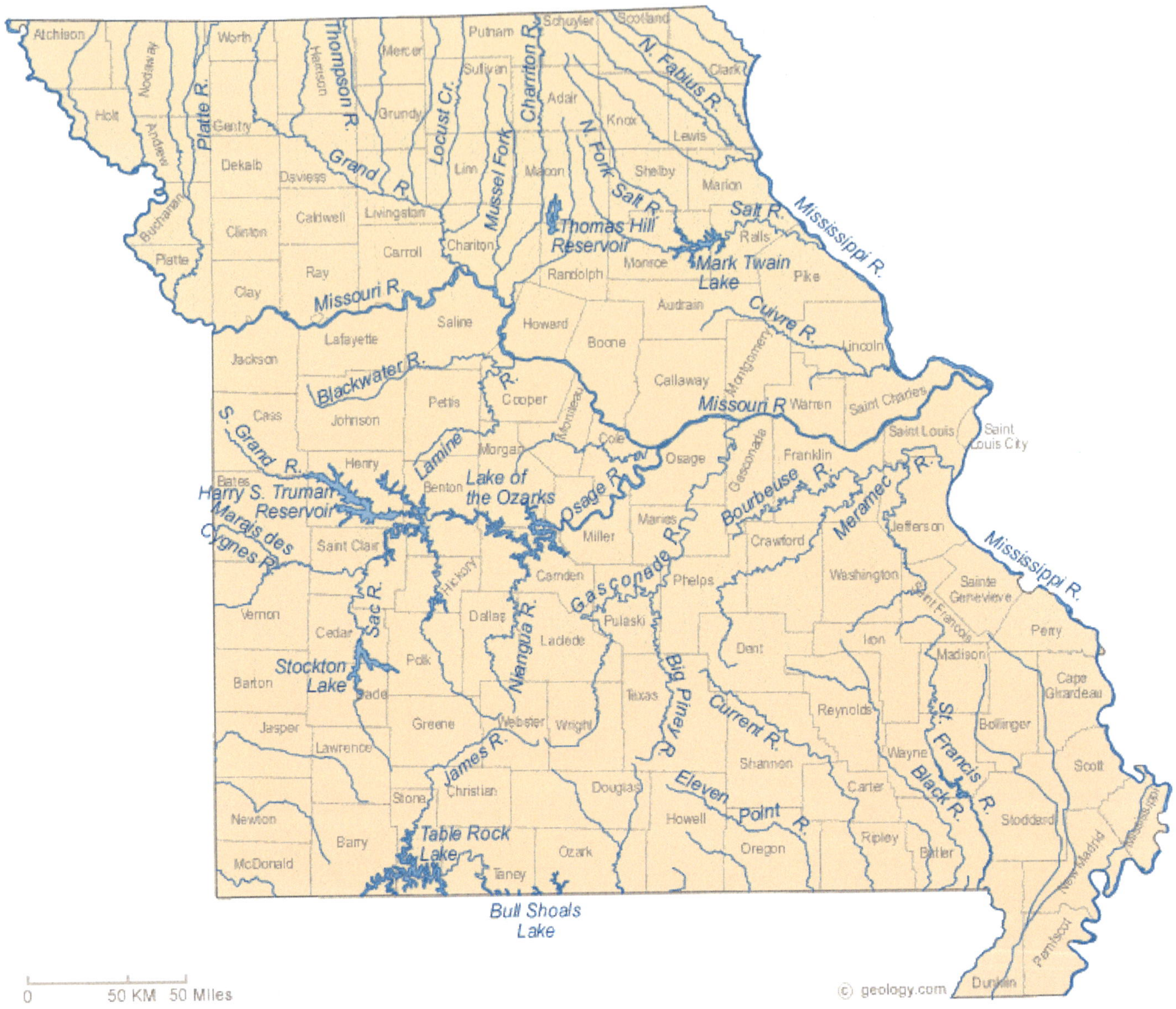

Rivers in Missouri

Image: Geology.com

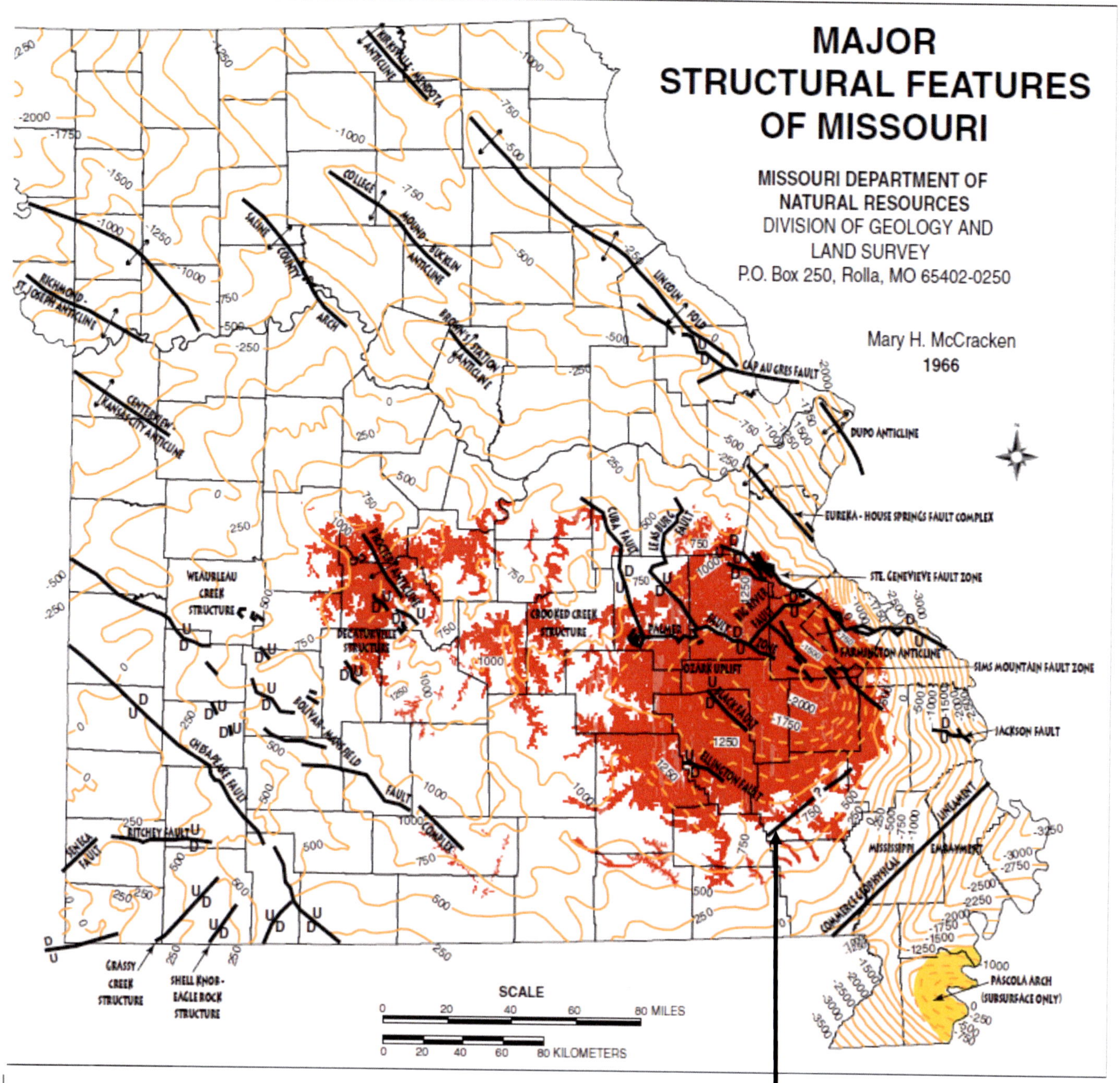

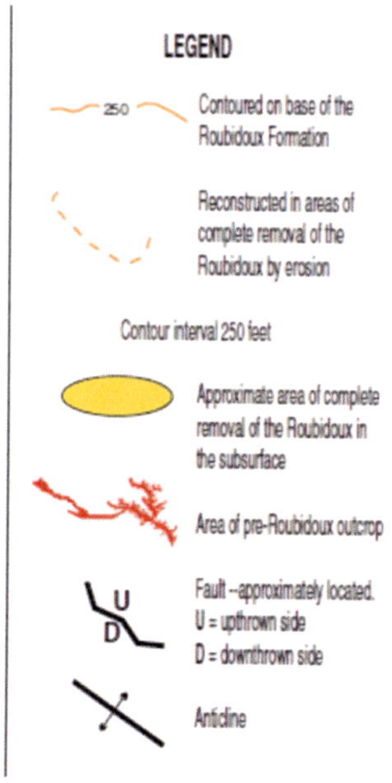

Wayne County has a fault going through the entire county

Fault lines in Missouri
Image: geocoaching.com

Note the fault line that goes through Wayne County. In my mind, this should be the primary focus of future investigations. Margie Kay

More Newspaper clippings from 1973

LOS ANGELES EVENING AND SUNDAY

HERALD EXPRESS EXAMINER

United Press International • Associated Press • Dow Jones

CLASSIFIED ADVERTISING Richmond 8-4111 All Other Calls Richmond 8-1212

VOL. CII NO.359 SATURDAY, MARCH 24, 1973 A PRICE 10 CENTS

CR: FRITZ KRON

UFO Following Cars in Missouri

Hovering Bright Light Reported by Hundreds

PIEDMONT, Mo. (UPI) — A UFO researcher said Friday the "Piedmont Unexplained," which hundreds of local residents have reported hovering over the Ozarks countryside or following cars down the highway, may be a mining expedition from outer space.

Another UFO was reported sighted in Illinois Thursday night.

"This could be the most significant, best-documented sighting in history," Hayden C. Hewes, head of the "International UFO Bureau" in Oklahoma City, said of the object seen in Piedmont.

Area residents, accustomed to such sightings since 1942, have remained calm since Feb. 21 when Piedmont High basketball coach Reggie Bone said a bright light followed a car carrying him and five players.

The object appeared on a desolate road between Piedmont and Ellsinore, Bone said. The light stopped and hovered over a field about 100 yards away before quickly rising out of sight.

Hewes said his UFO logging bureau has a photograph showing what looks like flying gyroscopes as well as reported cases of interference with television sets at the time of nearby sightings.

Another UFO was reported sighted in Grand Tower, Ill., Thursday night. An operating engineer said he saw an oval shaped object 25 feet in diameter with "red pulsating lights around the edge" hovering over a 66,000-volt transformer yard operated by Central Illinois Public Service Co.

"When it moved the lights would grow dimmer and then would brighten when it was still," said Oscar Wills, 57.

"I walked about 200 yards to the transformer yard and got almost to it before it moved away. It went right over me. It was about 1,500 feet up and was making no sound."

Wills said that after the object disappeared four jet aircraft which he presumed were military planes made 12 to 15 passes over the plant on sweeps into Illinois and Missouri as if they were looking for something.

UFO Sighting Stirs Missouri Community

PIEDMONT, Mo. UPI — There have been hundreds of reports of strange flashing lights in the sky around this countryside noted for its rich lead fields, and a UFO researcher thinks they are caused by a mining expedition from outer space.

"This could be the most significant, best-documented sighting in history," said UFO expert Hayden C. Hewes, who heads the International UFO Bureau in Oklahoma City.

Hundreds of people in this Ozarks community of 2,000 have reported seeing strange flashing lights, either high in the sky or at treetop level. Some say the objects that give off the light surfaced from nearby Clearwater Lake.

Reports of UFO sightings even came from Grand Tower, Ill., where Oscar Wills, an operating engineer at the Central Illinois Public Service Co. power generating station on the Mississippi River, reported Thursday an object hovering over a transformer.

"I was dumbfounded," said the 57-year-old Wills, who has worked at the plant 30 years. He said he didn't believe in unidentified flying objects until the sighting. Wills was described by the company as a "solid, dependable" employee.

The common description coming from Piedmont sightings is that the light is red, then turns multi-colored while spinning white, green, red and amber lights. It moves without a sound and is very fast.

Hewes said his UFO logging bureau has collected three types of evidence of the brightly lit red and amber UFO. He said there have been multiple witnesses, photographic evidence and electrical interference with television sets at the time of nearby sightings.

Hewes and two assistants visited with Piedmont residents last weekend. They interviewed and recorded the accounts of eye witnesses in hopes of supplying an answer.

The photographic evidence is supplied by Maude Jefferis, who teaches photography at Clearwater High School. A picture taken March 14 from the porch of her home shows an object like a child's top or gyroscope hanging in the night sky.

Reggie Bone, basketball coach at Piedmont High School, said a bright light followed the car in which he and five athletes were riding on a desolate road from Ellsinore to Piedmont. The light, Bone said, stopped and hovered over a field about 100 yards from the road. The object suddenly rose and disappeared at a high speed.

THE SOUTHEAST MISSOURIAN
CAPE GIRARDEAU, MO. THURSDAY, OCTOBER 4, 1973

Encounter with UFO frightening

By JOHN G. ADNEY
and
REGINALDE MITCHELL
Missourian staff writers

An encounter with an unidentified flying object near Cape Girardeau has left a Greenville truck driver shaken over what he claims he saw in the two to three minutes before unusual flashing lights disappeared.

Eddie Webb, 45, a driver for Sam Tanksley Trucking Inc. of Cape Girardeau, claims to have made contact, the hard way, with a UFO Today he is the center of attention as UFO sightings become more widespread throughout the area.

Mr. Webb was given emergency eye treatment at Southeast Missouri Hospital here and further treatment at a Cape Girardeau opthamologist's office

The State Highway Patrol is holding his twisted, nearly melted pair of eyeglasses as evidence that something unusual happened, although Mr. Webb is the only one to claim that he actually saw what he thinks is a UFO.

His wife, Velma Mae, also a truck driver and who was with him when the incident occurred, says she saw nothing.

And the doctor who treated Mr. Webb says he cannot confirm any type of injuries.

But Mr. Webb sticks to his story that he was temporarily blinded and burned. His wife believes him, and the highway patrol is puzzled.

The story he relates seems fantastic, almost unbelievable, but even the UFO skeptics who have seen his glasses' frames—twisted by a hot flash of unknown origin — admit there is something to that story.

GIVES ACCOUNT

Mr. Webb gave a detailed account of the incident to The Missourian Wednesday in a telephone conversation and later in an interview at his home in a secluded wooded area near Greenville in Wayne County. Mr. Webb, relaxing at home Wednesday night in a reclining chair and wearing sunglasses, appeared exhausted and shaken by his experience.

His forehead appeared slightly red, as if it had received a mild sunburn, and he complained frequently of pain which went deep inside his forehead and eyes. His sight had partially returned and he could see with blurred vision for about three feet, he said, but still flinched at bright lights.

He said his encounter with what he calls a turnip-shaped UFO with red and yellow lights in the center occurred between 6 15 and 6 30 a m. Wednesday on Interstate 55 just south of the Jackson exit at Wedekind Park He was driving a southbound Tanksley tractor-trailer rig Mrs. Webb, herself a truck driver for 16 years, was beside him in the cab when he first noticed in his rearview mirror the unusual lighted object behind the truck

Mr Webb said the time span from when he first saw the lights behind him until he was blinded by a bright flash was only two to three minutes

He and his wife had alternated driving from Parkersberg, W. Va , where they had picked up a load of raw plastic material packaged in 1,000-pound boxes. Mr. Webb had driven the last leg from Effingham, Ill , to the I-55 rest stop north of Fruitland, where they rested and left about 6 a m.

SEES LIGHTS

About 15 minutes later, just after they had passed the Jackson exit, Mr. Webb first noticed the lights of the object, about a mile behind him. He said from this far away all he noticed was the lights, only red and yellow, glittering and mixing together.

He said the lights were about four to five feet off the ground and at first he thought it was a flatbed trailer's lights or someone running without their headlights. He said he was driving his rig about 55 or 60 miles per hour and the lights appeared to be moving three to four times as fast as he was and were rapidly gaining on him.

"When I saw the lights, I told my wife that something was coming up behind us real fast," he said. "But when she looked out her mirror she said she didn't see anything at all

"I looked out my rearview mirror and saw them again, this time much closer" he continued

He said he saw an object which looked like a big turnip almost behind his truck and for an instant got a glimpse of it

"The bottom and top sections were spinning and looked like they were made from aluminum or chrome," he said in describing the object. "The center section had the lights on it and wasn't spinning, but the lights were glittering and the red and yellow seemed to mix together "

He said the object appeared to be about 30 feet in diameter, with parts of it covering the other lane of the road and the other part hidden behind his truck He said the lights were about four feet off the ground and that it must have been at least 10 feet in height.

HITS FOG

As he got this glimpse of the object almost on the tail of his truck and ready to hit him, the truck entered a patch of fog he said

Mr Webb stuck his head out the window and looked back toward the object "I saw something which looked like a light from a spotlight come down on the tandem of the trailer and at the same time the object started to rise higher in the air," he said.

He said up to the time the object caught up with his truck, he had heard no motor noise. But as it started to rise he noticed a "humming" sound which seemed to rise in frequency as the lights rose higher.

Mr. Webb said, "I stuck my head out a little farther looked back, and a bright white flash like a ball of fire hit me in the face. When the light flashed, everything stopped — the noise and all — and the lights went out "

Mrs. Webb said her husband after looking out of the window on the driver's side of the cab pulled his head back in and

UFO SIGHTINGS

screamed. He put his hands over his eyes and screamed that he couldn't see."

Mrs. Webb said she never saw any of the lights or heard the humming noise.

STOPS TRUCK

"At first I thought my truck lights had gone out so I put on the brakes and stopped in the middle of the road;" Mr. Webb said. "My wife told me I was cracking up and to get in the bunk in the cab and let her drive. It was then I realized I had gone blind."

Mrs. Webb turned a light on in the truck and looked at her husband to find his forehead red and hot as if it were blistered. One lens had fallen out of his glasses and the frame was melted and twisted.

After the truck was halted, Mrs. Webb took over and drove to the Tanksley complex on Highway 61 South, near municipal airport. She said they arrived about 6:45. An ambulance was called to the Tanksley office to take Mr. Webb to the hospital. He was given emergency treatment and a doctor examined him about 7:30. Mr. Webb was taken to the doctor's office about 8:30 where he remained most of the morning.

Mr. Webb's glasses have also been brought to the center of attention. Police who looked at the glasses said they didn't know whether the "melting and twisting" had been caused by heat or acid.

Sgt. Edwin E. Wright of the State Highway Patrol took the glasses to Poplar Bluff to have them sent to the patrol's laboratories in Jefferson City. He said, however, the glasses were to be returned to Cape Girardeau and tests would be run on them today by Dr. Harley C. Rutledge, head of the SEMO State University physics department, in an effort to determine what caused them to melt.

Dr. Rutledge has been involved during the past six months in a scientific investigation of mysterious lights and UFOs.

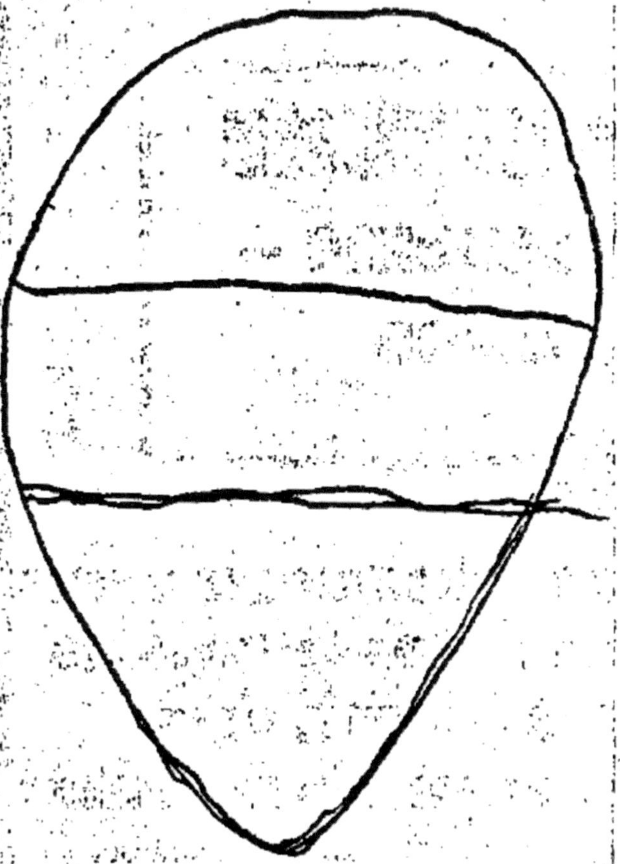

What he saw

Above is a sketch drawn by Mr. Webb of the unidentified flying object he observed. He said top and bottom sections were metallic and spinning, while center section was made up of red and yellow glittering lights.

The Cape Girardeau Southeast Missourian,

Thursday, October 4, 1973

D.O.S.

OCTOBER 3, 1973

(WEDNESDAY)

CONDUCTS TESTS

The Cape Girardeau doctor who examined Mr. Webb at the hospital and in his office said he ran several tests and made x-rays but could turn up no signs of burns, blindness or other injuries.

"I can confirm nothing," the opthamologist said.

Dr. John Englehart of Lutesville, the Webbs' family physician, theorized that it could have been some type of laser beam which caused the glasses to be distorted.

A policeman thought it might have been gas formed by the load of plastic in the trailer. But when the trailer was checked, there was no damage to the cargo or the truck, authorities said.

Whatever it was, it wasn't a helicopter or a motor vehicle, Mr. Webb said. He added that his wife saw nothing pass them, so that whatever it was must have gone up over the top of the trailer.

He admitted that he was frightened, first when the thought something was going to run into the back of the trailer and it didn't and again when everything went black, he said.

DON'T KNOW

"As far as UFOs and flying saucers are concerned, I always figured that someone who saw them either was pilled up or had been drinking," Mr. Webb said. "But I don't do either and I saw something unexplainable. So now I just don't know. I'm almost afraid to think of what I saw. I'd think someone might put me in a straitjacket."

Despite the incident, Mrs. Webb says she still does not believe in UFOs. "I never saw the lights; I didn't see anything," she said. "But I heard my husband scream and I saw his glasses. I know he was frightened and that he was in pain. I saw burns on his forehead and around his eyes.

"I was really scared," she continued. "But I still don't believe in UFOs. I don't think there have ever been any."

Sighter's vision returns

DALLAS TIMES HERALD
OCT. 5, 1973

Glasses carry UFO's mark

CAPE GIRARDEAU, Mo (UPI) — A physics professor investigating a reported attack on a truck driver by an unidentified flying object (UFO) said today the victim's eyeglassese were damaged by internal heat from an undetermined source.

In the latest of a series of recent UFO sightings in southeastern Missouri, Eddie Doyle Webb, 45, of Greenville, Mo., was blinded for several hours after the incident. He is recovering his vision but intends to visit an eye specialist at Barnes Hospital, St. Louis.

Webb said he was driving a tractor-trailer rig about dawn Wednesday when he saw a bright object in the air behind him, "coming up real fast."

He awakened his wife, Velma Mae Webb, 47, who was asleep in the cab, he said, but she didn't see anything.

"Then, I stuck my head out of the window and a large ball of fire struck me in the face," Webb said. "My glasses fell off and I couldn't see. But I got the truck stopped."

Mrs. Webb said her husband screamed, "Oh, my God! I'm burned! I can't see!"

One of the lenses of his glasses fell out of the plastic frame, which was warped. Mrs. Webb, who serves as a relief driver at times, drove him to a hospital.

Sgt. Ed Wright of the Highway Patrol took Webb's glasses to Dr. Harley Rutledge, head of the Southwest Missouri State University physics department, for an analysis.

Rutledge, who has been working for six months in an attempt to identify mysterious flying objects, said he put the glasses under a microscope and "it appeared they were heated internally.

"The plastic apparently got hot and the mold came to the surface. The heat warped the plastic, causing the lens to fall out."

Rutledge said he planned more tests on the glasses. He said there appears to be "some residue which we hope to put through some chemical tests."

Scientists say huge 8-year study proves UFOs are real

By TONI REINHOLD

A RESPECTED physicist has presented the most convincing case for the existence of UFOs ever documented in America after an eight-year study in which he and a team of scientists made 178 sightings of mysterious flying objects.

Focus of the study was the small Missouri town of Piedmont, 160 miles south of St. Louis, which has become known as the UFO capital of America because of the large number of sightings in its vicinity.

Dr. Harley D. Rutledge, chairman of the physics department at Southeast Missouri University, says in his sober, low-key appraisal of the UFO question: "According to known technology, many of the UFOs were not man-made."

Dr. Rutledge, 55, who has been with the university's physics department for 17 years, put his career and standing in the academic community on the line when he launched his massive investigation of UFOs in 1973.

Its culmination is a book, Project Indentification – The First Scientific Field Study Of UFO Phenomena (Prentice-Hall, $10.95), to be published next month.

Dr. Rutledge says his scientific curiosity was aroused by the Piedmont sightings and adds: "It seemed to me the UFO phenomena wasn't being approached logically.

"I was also concerned about the effect the reported sightings were having on the people of Piedmont. Women were so frightened that they were carrying firearms."

James D. Sutton, Piedmont's mayor, recalls: "There were definite patterns and times people were seeing them. I saw one myself. It was a huge ball of bright light moving under heavy clouds. No plane flies that low under cloud cover."

Dr. Rutledge and his team were well equipped to tackle the job scientifically, and nearly 40 scientists, engineers, students and others went into the field at various times.

Of all the 178 sightings, Dr. Rutledge says he was most impressed by two encounters – one at night, the other during the day.

Recalling the daylight sighting, he says: "I saw a dull gray object. I did a double take – it was no airplane. I did another double take. It had no wings. It was shaped like a bullet. There was no sound coming from it. The body was untapered and the rear was flat.

"I took my eyes off it for a second to pick up my binoculars and it was gone.

"Of all the sightings I had seen up to that point, that one upset me quite a bit. I felt nauseous. I couldn't kid myself any longer about UFOs. I had to accept them."

During an aerial search Dr. Rutledge made the sighting that excited him most. "We were flying at about 3,500 ft. and I was scanning the sky with my binoculars when I saw a light. It began to pulse, about once a second. I told the people in the plane that we had a friend up there.

"Then a second pulsing light came on to the lower right. The pulsations of the lights didn't look like a simple form of communication. Then a row of seven lights came on between the first two. Then another light came on. I witnessed an incredible thing happening in the sky.

Dr. Harley Rutledge with the special instrument he used to measure the height of a UFO.

★

Dr. Rutledge and his team made most of their sightings in the vicinity of the small Missouri town of Piedmont, which has become known as the UFO capital of America.

"The seven aligned lights were not identical. Three of the lights on the right seemed closer and were surrounded by what appeared to be a cloudlike, shimmering haze. The four other lights were farther away and appeared to twinkle and vibrate in position.

"I gave my binoculars to someone else in the plane, but they couldn't locate the lights. I grabbed back the binoculars but the lights had vanished."

Summing up his UFO experiences, Dr. Rutledge says: "I have not yet been able to determine where the objects came from, but based on the things I saw it is difficult to accept that our world has the technology to produce them."

From the pageant headquarters in Biloxi, Miss., Glasser said: "If we find a girl wearing falsies she is ruled out of the contest. Miss Fountain did. The rules are quite explicit.

"As for her allegations, we closely watch our contestants and we have found no irregularities. If a girl has had silicone help, she is banned."

Firemen get hot under collar over a horse

A HORSE got some firemen hot under the collar – because he kept pulling false alarms.

The firemen, in Del Mar, Calif., rushed to the Del Mar National Horse Show grounds to answer an alarm. But there was no fire and nobody around the fire box when they arrived.

While they were standing around trying to figure out who set off the alarm, a horse named Jonathan trotted up and pulled it again, right before their eyes.

Jonathan's owner, Pam Johnson, of Mira Loma, Calif., says the horse is full of little tricks like that.

GEORGE WASHINGTON ANNIVERSARY COINS

GEORGE WASHINGTON's 250th birthday will be celebrated next year with the minting of $100 million in collectors' coins. The Treasury Department will mint 10 million Washington half-dollars, each of them 90 percent silver. Collectors are expected to pay $10 apiece for the non-circulation coins.

★★★★★★★★★★★★★★★★★★★

THERE is only one thing worse than being talked about, and that is not being talked about. – ***Playwright Oscar Wilde.***

DAVID A. SCHROTH
3517 Pennsylvania Avenue
St Louis, MO. 63111

WAYNE COUNTY JOURNAL–BANNER
Thursday, April 12, 1973 Page 2

UFO Landing Site Or ?

Several trees have been found damaged just off the highway four miles west of Ellsinore on Hwy. 60. In addition to trees being damaged three holes in the ground were found in a triangular formation, reports Mrs. Stella Harlow of Piedmont, who with her father Cletis Richmond, investigated the area and took the pictures showing the damaged trees.

Three trees were damaged beside where the three holes were found, Mrs. Harlow reports, and each of the three trees are pictured. She said the three holes were about 6" deep and 2 to 3 feet across.

One of the three trees was split down the center and splintered to pieces, another tree had the bark stripped down one side of it about 40 feet high, a third tree in the triangle had only a little bark torn off at the trunk next to the hole in the ground.

Several other trees were damaged within the triangle which Mrs. Harlow estimates to be about 50 feet between points.

Mrs. Harlow said a man about 70 years old was also observing the area when she was there Monday afternoon and he said that he had seen lightning damage several times during his life but that he had never seen lightning do anything like this.

Mrs. Raymond Stucker lives across the highway from the damaged trees and reports she saw an object with what looked like three landing feet above the tree line Friday, April 6, about 11:30 a.m. Mrs. Stucker made a sketch of what she saw for Hayden Hewes, director of the International UFO Bureau of Oklahoma City, Oklahoma.

CLETIS RICHMOND stands by a tree which was splintered to pieces.

STELLA HARLOW looks at a tree which has the bark stripped off one side of it.

AARP To Have UFO Discussion

The next regular meeting of A.A.R.P. will be at the Greenville Methodist Church on April 16, at 2:00 p.m. This is to be

Article continued from previous page:

WAYNE COUNTY
Thursday, April

Vol. 97, No. 8

a sketch of what she saw for Hayden Hewes, director of the International UFO Bureau of Oklahoma City, Oklahoma.

AARP To Have UFO Discussion

The next regular meeting of A.A.R.P. will be at the Greenville Methodist Church on April 16, at 2:00 p.m. This is to be a very important program on the current UFO, now taking place in our area. If you want to hear all the latest data collected, see all the evidence and ask all the questions you like, please be there and the group working on trying to solve the problems now in question will answer to the best of their ability, the things you would like to know. There are some things we can not answer, because we do not know, but there are some things that you, as Older Citizens, should be told. Get a car load from your area, and be at Greenville on April 16, at 2:00 p.m., at the Methodist Church. Always an interesting meeting.

STELLA HARLOW looks at a tree which has the bark stripped off one side of it.

Open Forum

Editor, Ledger:

Look to the southern sky any clear night and your chances of seeing an actual UFO are very good because the 1974 summer flap is in full scale action. Flap is UFO-ologist language for a high rate of UFO activity. The secret for seeing a UFO is knowing what to look for. From Mexico it is much easier to see a UFO AT night than in the daytime because UFOs seem to congregate around geological fault lines and the closest fault line to Mexico is in the Ozark region particularly around the Farmington and Cape Girardeau area.

UFO normally appear at night as a bright light, which means a UFO at high altitude is visible in that area from Mexico. It takes a little patience and sometimes your efforts are not rewarded, but they are out there. I've seen two this week. What they normally look like is a bright star that is moving at a high rate of speed and could be mistaken for a meteor or falling star except falling stars go toward the horizon. However, UFOs either come up from the horizon or parallel to it. Sometimes they can be mistaken for an airplane, but airplanes do not stop in flight, reverse direction without turning around, or disappear in mid flight.

If you see a bright light traveling at a high rate of speed that does some of these things you have seen a UFO. It isn't an airplane or falling star. It is an unidentified flying object and that is what a UFO is.

Carl B. Thompson

Wayne County Wonders What It Is Up In Sky

PIEDMONT, Mo. (AP) — "There's no doubt there's something up there. We just don't know what it is," said Gene Beardon, police chief in this southern Missouri town, Tuesday.

Beardon was talking about a redish-orange light that he said he and about 500 other people have seen "flying" around the skies of Wayne County in the last month.

The phenomenon has been called by many Piedmont residents a UFO—Unidentified Flying Object. It was first seen Feb. 21 by Reggie Bone, Clearwater High School basketball coach, and five of his players.

Bone claimed he and the boys saw the object flying about 50 feet above the ground along Brushy Creek, 15 miles south of town, as they were driving back from a basketball tournament at Essex, Mo.

Since then, the UFO has been sighted numerous times, Beardon said, usually on "dark, clear nights." He said it has normally been seen flying at a high rate of speed and turning and jumping from side to side.

The chief said, however, when he saw the light in his yard, it was moving slowly and even standing still at times. He said it did not make any noise.

Maude Jefferies, a high school photography teacher, said she managed to take some pictures of the UFO on March 14. She said from the photographs, which were taken at least a mile from the light, the object looked like a gyrotop.

She said she could not tell whether it was solid, but it appeared to be self propelled.

Beardon said the UFO has stirred up the interest of persons outside the town of 3,000. He said last weekend Harlan Hewes, president of the International UFO Society, was in Piedmont interviewing persons who had seen the light.

"We don't know if it's a government experiment or what," Beardon said. "But there's definitely something here. It's not an illusion."

He said no damage or injuries have been reported as a result of the UFO, and "for the time being, we're just observing it."

Former Judge Dies

ROLLA, Mo. (AP) — William E. Martin, former Republican state legislator and Crawford County Judge, died here Tuesday. He was 83.

Martin represented Crawford County in the Missouri House from 1938 to 1944 and later served as county judge for 12 years.

Service will be Thursday in Cuba, Mo.

March 29, 1973

Photographs of Pie

PUBLIC NOTICES

NOTICE OF FILING
OF FINAL SETTLEMENT
AND PETITION
FOR DISTRIBUTION
STATE OF MISSOURI,)
COUNTY OF TEXAS)ss.
IN THE PROBATE COURT
OF TEXAS COUNTY, MISSOURI.
In the Estate of EARNEST STARK, Deceased.
Estate No. 4497

To all persons interested in the estate of Earnest Stark, deceased.

Notice is hereby given that a final settlement and petition for determinination of the persons who are the successors in interest to the personal and real property of the decedent and of the extent and character of their interests therein and for distribution of such property, will be filed in the Probate Court of Texas County, Missouri, by the undersigned on the 1st day of May, 1973, or as continued by the court, and that any objections or exceptions to such settlement or peititon or any item thereof must be in writing and filed within ten days after the filing of such settlement.

Ella Stark Smith, Administratrix, Raymondville, Missouri, Address.
WILLIAM E. GLADDEN, Attorney, Houston, Missouri 65483, Address. Telephone Number 967-3520. 40/4t

* * *

NOTICE OF LETTERS
OF ADMINISTRATION
GRANTED
STATE OF MISSOURI,)
COUNTY OF Texas)ss.
IN THE PROBATE COURT
OF Texas COUNTY, MISSOURI
In the Estate of Marvin James Bailey, Deceased.
Estate No. 4626

To all persons interested in the estate of Marvin James Bailey, decedent:

On the 23rd day of March, 1973, Norville T. Hammon appointed the administrator of the estate of Marvin James Bailey decedent, by the Probate Court of Texas County, Missouri. The business address of the administrator is Star Route, Raymondville, Missouri whose telephone number is 457-6420 and his attorney is Maurice W. Covert whose business address is Houston, Missouri, and whose telephone number is 967-2433.

All creditors of said decedent are notified to file claims in court within six months from the date of this notice or be forever barred.

All persons interested are notified that such court, at the times and as provided by law, will determine the persons who are the successors in interest to the personal and real property of the decedent and of the extent and character of their interests therein.

Date of first publication is March 29, 1973.

Dorothy S. Akins, Clerk, Probate Court of Texas County, Missouri. 40/4t

Men In Service

Two Texas County men recently enlisted into the U. S. Army in ceremonies at St. Louis.

William W. Connolly, Raymondville, selected the Army's two year enlistment option and is now assigned to the Army Center, Fort Leonard Wood. Connolly has received his B. S. degree in Physics from Washington University and has completed part of his credits for a master's degree. He plans to complete the required courses for a master's either while on active duty or by utilizing the GI bill upon completion of active duty.

Rolla E. McCoy, son of Mr. and Mrs. Elmer McCoy, Houston, enlisted into the Army's six-month delay entry program and has been assigned to Fort Wood for eight weeks of transitional training.

Mary S Chapter of OES To Meet

Mary S Chapter #126, O E S will meet for regular meeting, Tuesday, April 3 at 8 p.m. Members are urged to attend. There will be a Memorial for deceased cancer victims and we will be honoring all Past Matrons and Past Patrons.

Blanche Kyser, W.M.
Willis Willson, Secy.

dmont UFO

These photographs appeared in the March 22 edition of the Wayne County Journal-Banner at Piedmont, Mo. and were taken by Maude Jefferis about 11:30 p.m. on March 14 with a Crown Graphic 4x5 camera mounted on a tripod with an exposure time of 10 minutes. The white streaks in the sky are stars and show the distance they traveled during the 10 minute exposure.

The arrow points to the UFO. At left is an enlarged photograph of the UFO designated by the arrow. The picture at left was made from the same negative as the one used to produce the photo at right.

The Piedmont newspaper was brought to the office of The Houston Newspapers by Mr. and Mrs. Sam Ijames, Houstonians who formerly lived at Piedmont.

On a recent evening in Houston, various persons telephoned various other persons with the same message:

"Go outside and look in the sky to the East. It's something unusual. Could it be a flying saucer (UFO)?"

What several persons reported seeing might have been what hundreds have apparently been sighting around Piedmont, Mo., some 100 air miles east of Houston.

Some reported the UFO had flashing, colored lights. Others said it was just a bright light. A few said they saw it move away at high speed.

The Piedmont "thing" has been reported widely with wire services carrying stories. St. Louis newspapers have sent reporters to the scene. Tv and radio news reports have added impetus to the story.

Piedmont residents visiting in Houston say that hundreds of persons have been converging on the small town, literally causing traffic jams.

The UFO reportedly had not been seen for several days prior to Monday.

The following news story was distributed Saturday by UPI:

PIEDMONT, Mo. (UPI) – A researcher in unidentified flying objects says the "Piedmont Unexplained" may be on a mining mission to this lead-rich area of the southern Missouri Ozarks.

Hayden C. Hewes, head of the International UFO Bureau in Oklahoma City, said Friday he is hopeful the Piedmont sightings will provide him his first closeup look at a UFO after investigating the phenomenon for the last 15 years.

Hundreds of area residents say they have seen disc-shaped objects, though descriptions have been varied. Some say the objects surfaced from the reservoir side at Clearwater Dam. The objects turned from red to multi-colored spinning white, green, red and amber lights.

* * *

Residents of the area, accustomed to such sightings since 1942, have remained calm the weeks since the high school basketball coach and five athletes reported the first sighting the night of Feb. 21.

Reggie Bone, coach at Piedmont High School, said a bright light followed the car in which he and the athletes were riding a lonely road from Ellsinore to Piedmont. The light stopped and hovered over a field about 100 yards from the road. The object suddenly rose and disappeared at a high speed.

Since then Piedmont Police Chief Gene Bearden has received hundreds of reports.

"My first impression was that it is something McDonnell Douglas or the government is experimenting with," Bearden said. "Obviously they haven't harmed anyone and don't seem to pose a threat.'

* * *

One Piedmont resident, Omer Fears, said he thinks the bobbing lights may be marsh gas or an experimental device, but added, 'I'm from Missouri and I'm going to have to see it before I believe it."

Hewes said there are fields of nearly pure lead in the area and the UFO may be on a mining mission.

'I think there's a base of operation at the lake," he said. "We've had eight persons report seeing it in or coming out of the water. I don't know what the connection is, but we have had too many sightings near water for it to be coincidence."

* * *

Hewes said his UFO logging bureau has collected three types of evidence of the brightly lit red and amber UFO – multiple witnesses, photographic evidence from high school photo teacher Maude Fefferis and electrical interference with television sets at the time of nearby sightings.

"This could be the most significant, best-documented sighting in history," Hewes said.

Mrs. Jefferis claims to have taken time exposure photographs of several flying objects from the porch of her home behind a drive-in theater. The pictures, taken March 14, show an object shaped like a child's top or gyroscope hanging in the western sky.

Mrs. Jefferis said she has seen as many as four objects in the sky at one time.

The strange lights over Missouri

One scientist's close encounters on his mission to unravel the mysteries of the sky. **Features**

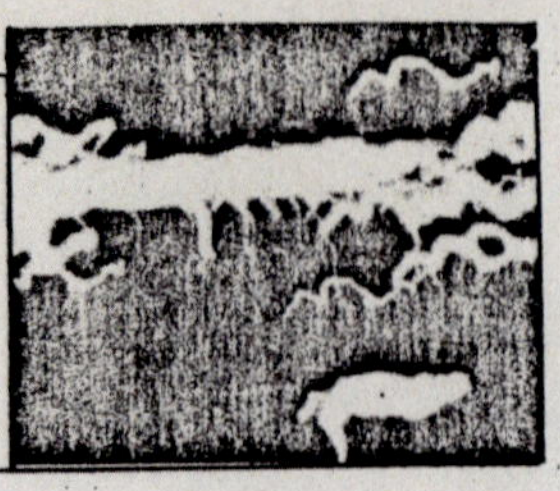

Saturday — Sunday
July 11- 12, 1981
50¢
Section A — Pages 1-12

St. Louis Globe-Democrat

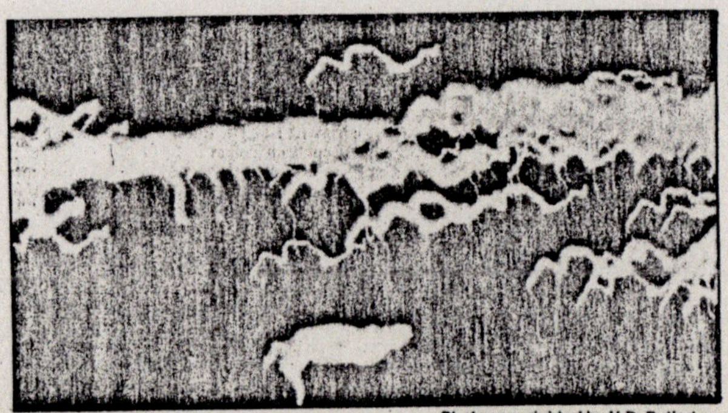

Photo copyrighted by H.D. Rutledge

Famous photograph of mysterious lights as they glided silently over the Farmington, Mo., airport in May of 1973.

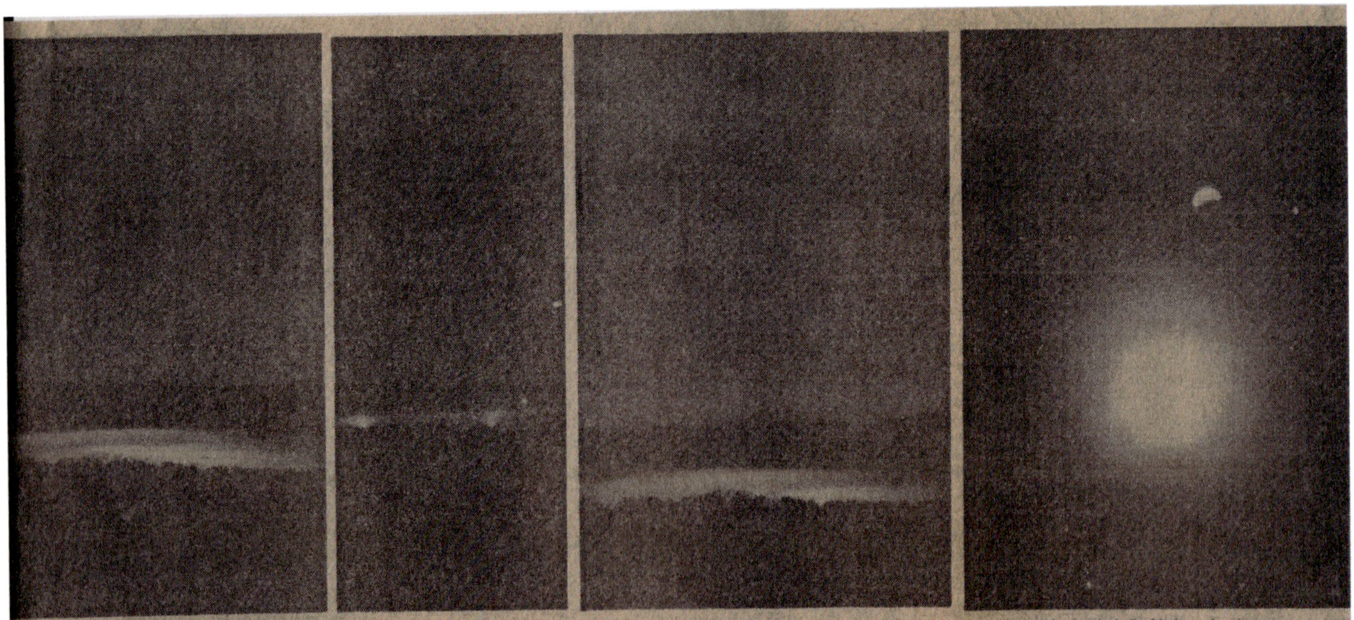

Suddenly, a glow. | Points of colored light in the sky. | The glow again. | And a bright ball of light under the moon.

Photos by Ben Ba

Piedmont Aerial Carnival Turns A Whole Family Into UFO Believers

Piedmont, Missouri 63957 – Thursday, March 29, 1973

WAYNE COUNTY JOURNAL BANNER

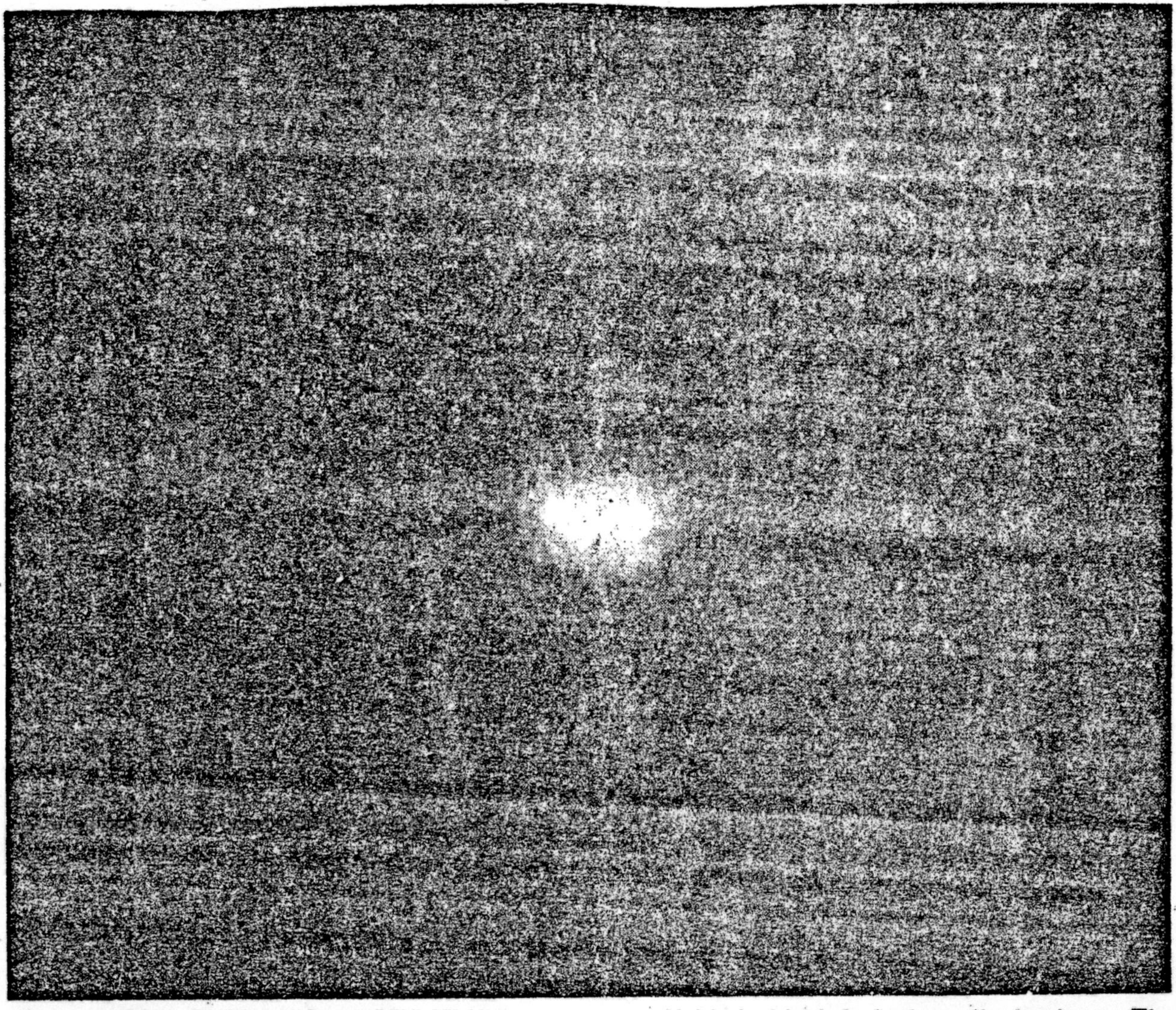

GARY SUTTON'S PHOTOGRAPH pictures a lighted object just above the horizon. The photograph was taken from a gravel road behind Worley's Sawmill near Piedmont facing in the direction of Clearwater Lake.

UFO blinds motorist?

CAPE GIRARDEAU, Mo. (UPI) — A physics professor investigating a reported attack on a truck driver by an unidentified flying object (UFO) said today the victim's eyeglasses were damaged by internal heat from an unknown source.

In the latest of a series of recent UFO sightings in southeastern Missouri, Eddie Doyle Webb, 45, of Greenville, Mo., was blinded for several hours after the incident. He is recovering his vision, but intends to visit an eye specialist at Barnes Hospital, St. Louis.

Webb said he was driving a tractor-trailer rig about dawn Wednesday when he saw a bright light or aluminum object in the air behind him, "coming up real fast."

He awakened his wife, Velma Mae Webb, 47, who was asleep in the cab, he said, but she didn't see anything.

"Then, I stuck my head out of the window and a large ball of fire struck me in the face," Webb said. "My glasses fell off and I couldn't see. But I got the truck stopped."

Mrs. Webb said her husband screamed, "Oh, my God! I'm burned! I can't see!"

One of the lenses of his glasses fell out of the plastic frame which was warped. Mrs. Webb, who serves as a relief driver at times, drove him to a hospital.

Sgt. Ed Wright of the Highway Patrol took Webb's glasses to Dr. Harley Rutledge, head of the Southwest Missouri State University physics department, for an analysis.

Rutledge, who has been working for six months to attempt to identify mysterious flying objects, said he put the glasses under a microscope and "it appeared they were heated internally.

"The plastic apparently got hot and the mold came to the surface. The heat warped the plastic, causing the lens to fall out."

Wash. State.

2A Everett Herald, Friday, October 5, 1973

About Missouri MUFON and How to File a Sighting Report

Missouri MUFON is a state chapter of the Mutual UFO Network. MUFON is a non-profit organization which had its beginnings in St. Louis, Missouri with the UFO Study Group in 1969, after Project Blue Book was disbanded by the government. MUFON is a civilian UFO reporting and investigative organization with chapters in over 40 countries. It is the largest such organization in the world.

After a report is made to MUFON at **www.mufon.com,** an investigator is assigned to the case. It does not matter when the sighting occurred, we want to have the information in our archives for research purposes, However, it is important to file a report immediately after a sighting while it is still fresh in your mind. An investigator will contact the witness and may have further questions. Sighting reports may be anonymous.

The Missouri State Director is Debbie Ziegelmeyer, and the Assistant State Director is Margie Kay. Our Chief investigator is Mel Van Vickle, and our Chief Laboratory Analyst is Lynne Mann.

Missouri has the only MUFON Lab in the world. It is located in Imperial, Missouri and serves all of MUFON.

We have 19 investigators across the state, but can always use more. If you are interested in becoming a field investigator please find the information on the MUFON website at www.mufon.com. After passing the test, the candidate becomes an investigator trainee, and after showing skill in doing investigations per the MUFON guidelines, they will be promoted to Certified Field Investigator status.

Missouri is a very busy state when it comes to UFO sightings, and we are consistently in the top 10 for UFO reports each month.

Our chapter is very active and holds monthly meetings and presentations by guest speakers on Zoom. This meeting is open to the public. To receive our Midwest MUFON newsletter and announcement list send an email to **momufonasd@gmail.com**.

In-person meetings are held monthly in Webster Groves (St Louis area) and quarterly or more often in the Kansas City area. Other chapter meetings may open soon.

We hope you'll join us.

Join MUFON at www.mufon.com

References

Cape Girardeau Daily Webb
Cattle-Killing UFO of 1896 Fate Magazine
cityofpiedmont.com

Project Identification: The First Scientific Field Study of UFO Phenomenon, by Harley Rutledge, PhD 1981

MUFON archives

MUFON Case Management System

Missouri Great Outdoors

Missouri-Vacations.com (native Americans)

"People are Saying" article, volume 32 #17 (22 December 1977)

Piedmont Chamber of Commerce

Poplar Bluff American Republic Newspaper

SKYLOOK

The Missourian

Top Scientist Reveals... I've Made 25 Sightings of Intelligently Controlled UFOs By Bernie Ward

Facebook group Piedmont UFO Watch

Missouri Minerals - Virtual Museum of Geology
Newspapers.com

From Wikipedia:
In 1966, Harley Rutledge completed his Ph.D. in solid state physics at the University of Missouri. He subsequently took the position of Professor and Dr. Harley Rutledge, chairman of the Physics Department at Southeast Missouri State University. He was Department Chairman from 1964 to 1982 and retired from teaching in 1992.

UFO Research
Challenged to explain sightings of unidentified lights and luminous phenomena in the sky around Piedmont, Missouri, Dr. Harley Rutledge decided to subject these reports to scientific analysis. He put together a team of observers with college training in the physical sciences, including a large array of equipment: RF spectrum analyzers, Questar telescopes, low-high frequency audio detectors, electromagnetic frequency analyzer, cameras, and a galvanometer to measure variations in the Earth's gravitational field.

The resulting Project Identification commenced in April 1973, logging several hundred hours of observation time. This was the first UFO scientific field study, able to monitor the phenomena in real-time, enabling Rutledge to calculate the objects' actual velocity, course, position, distance, and size.

Observation of the unclouded night sky often revealed "pseudo stars" - stationary lights camouflaged by familiar constellations. Some objects appeared to mimic the appearance of known aircraft; others violated the laws of physics. The most startling discovery was that on at least 32 recorded occasions, the movement of the lights synchronized with actions of the observers. They appeared to respond to a light being switched on and off, and to verbal or radio messages. The final results of this project were documented in the 1981 book, Project Identification: The first Scientific Study of UFO Phenomena.
Harley Rutledge, 80, former chairman of the physics department at Southeast Missouri State University, died on Monday, June 5, 2006 at the Missouri Veterans Home.

Books
Project Identification: The first Scientific Study of

UFO Phenomena. Prentice-Hall 1981 ISBN 0-13-730713-6 by Harley D. Rutledge, Ph.D.
Research Papers
Project Identification: Thirteen Years and One-Hundred and Sixty Sightings Later, Harley D. Rutledge, Ph.D. (presented at the 1986 MUFON symposium).
References

1. ^ Greaney, TJ (June 6, 2006). "Area UFO expert Dr. Harley Rutledge, 80, dies". semissourian. http://www.semissourian.com/story/1155552.html. Retrieved 2007-02-26.

Cape Girardeau Daily Webb
Cattle-Killing UFO of 1896 Fate Magazine
cityofpiedmont.com

Project Identification: The First Scientific Field Study of UFO Phenomenon, by Harley Rutledge, PhD 1981

MUFON archives

MUFON Case Management System

Missouri Great Outdoors

Missouri-Vacations.com (native Americans)

"People are Saying" article, volume 32 #17 (22 December 1977)

Piedmont Chamber of Commerce

Poplar Bluff American Republic Newspaper

SKYLOOK

The Missourian

Top Scientist Reveals... I've Made 25 Sightings of Intelligently Controlled UFOs By Bernie Ward

Facebook group Piedmont UFO Watch

Missouri Minerals - Virtual Museum of Geology
Newspapers.com

Recent article: Buck Nelson's Spacecraft Convention - Cape Girardeau History and Photos (capecentralhigh.com)

Recent article: The unsolved mystery of Buck Nelson: What happened to the Ozarks' first alien abductee? | The Focus Insider | phelpscountyfocus.com

Wikipedia: https://en.wikipedia.org/wiki/Buck_Nelson

The Hutchinson News: www.hutchnews.com/sotry/news/2022/07/01geneseo-museum-ufo-day-capital-kansas-saturday/7776407001/

The Geneseo City Museum, Geneseo, Kansas

www.nashvillescene.com/

www.metv.com

www.ashlandsource.com/

science.howstuffworks.com/

www.thestarpress.com

https://richlandcountyhistory.com/2019/09/02/ufos-over-richland-county-1973/

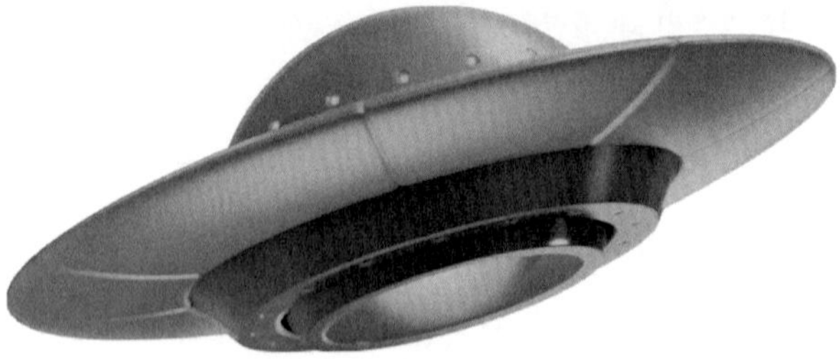

What to do if you see a UFO/UAP

1. Be prepared. Go into the settings on your cell phone camera and change it to the highest setting for video. This will switch it to at least 60 fps (frames per second) and provide more detail than standard video. Keep the phone with you at all times just in case!
2. Don't panic if you see a UFO.
3. Get our your cell phone and put it on VIDEO.
4. Turn the phone sideways (landscape).
5. Lean against something like a car or tree while filming to keep steady. Keep the UFO in the frame.
6. Try to keep something else in the video for reference like a tree, house, telephone pole, etc. if possible.
7. Keep filming until the UFO goes out of sight.
8. Check the GPS on your phone and write down your exact position. Also check the compass on your phone and write down the exact degree where you were looking where you saw the object.
9. Download an Inclinometer app on your phone and use it to check the degrees (how far up in the sky) where the UFO was located.
10. Write down the time and date, and all details you can remember such as sound, appearance of the object, odd animal behavior, car problems, physical reactions, etc.
11. If a landing is involved, stay away from the area and keep pets and children away.
12. File a report at www.mufon.com as soon as possible.

Compass heading is crucial to an investigation. Use a regular compass or download a compass app on your phone. There are 360 degrees on a compass and noting where a UFO was in relation to your position can provide information needed in order to rule out any standard type of craft.

Degrees are marked on the outside perimeter

Note: The police, sheriff, Highway Patrol, FAA, Air Force and Military bases do not investigate civilian UFO sightings. MUFON has investigators in all states and multiple countries and will have an investigator contact you.

Find us on: facebook®

Join us at

Piedmont UFO Watch

where people share videos and photos of UFOs in Wayne County

and

Like the **Missouri MUFON** Page

on Facebook to get updates about meetings and events

Also:

Missouri UFO/UAP and Weird Stuff Hot Spot group

covers the entire state

Debbie Ziegelmeyer of Imperial, Missouri, was born in Grand Rapids, Michigan, attended school in Safford, Arizona and later attended high school first in Southern California, and then Arnold, Missouri where she graduated from Fox Senior High School.

Debbie joined MUFON in 2000 and became a Field Investigator shorty after. She is currently a MUFON Star Team Investigator, the State Director of Missouri MUFON, a member of the MUFON Business Board of Directors, The Functional Director of Underwater Research and Recovery, a Benefactor, and Inner Circle member, and a member of the MUFON MARRS project which involves archiving of historic UFO file reports. She is also a member of Ted Phillip's S.I.U./T.P.R.C. Team, and in 2004, organized the MUFON Dive Team along with Tom Ferrario current T.P.R.C. Team leader. Debbie has over 1200 completed UFO investigations, with over 900 of them being MUFON sighting reports. She also teaches a MUFON Field Investigator Dive Team course which instructs investigators in both in water and on shore responsibilities and protocols. Debbie is also co-Founder along with MUFON State Section Director Lynne Mann, of the MUFON Analytical Research Laboratory which is located in southeastern Missouri.

Debbie has been a Roswell 1947 crash investigator for over 25 years, has been an annual Roswell, NM July Festival speaker, and has made several speaking appearances at the "UFO Watch Tower" in Hooper, Colorado. She was a volunteer archeologist on two NBC projects, 2002 "Roswell Crash Starling New Evidence", which published her personal experiences in the book "Roswell Dig Diaries", and the 2006 UFO Hunters episode "Roswell Revisited". She appeared on an episode of "Close Encounters" as an expert on the "1970s UFO Flap in Piedmont, Missouri", and was an expert consultant on a San Luis Valley episode of "Mysteries in National Parks" on the Travel Channel.

Debbie was a 2017 Devils Tower "UFO Rendezvous" speaker, a 2018 "Alien Cosmic Expo" speaker in Toronto, Canada, a 2013, 2017 and 2021 MUFON Symposium speaker, a 2019 MUFON Cruise speaker, an Arkansas Paranormal Expo speaker in 2018, 2019, and 2022 in Little Rock Arkansas, a 2021 OZ Conference speaker and a 2022 Shag Harbor UFO Expo speaker. She has also made several personal speaking appearances in Missouri and Illinois, has been a guest SKYPE and ZOOM speaker, and been a guest on several radio shows and podcasts including Un– X News.

Most recently, Debbie appeared on two episodes of "Alien Highway" as a consultant, two episodes of "Alaska Triangle", and the 2021 Discovery+ series "Roswell the Final Verdict". Debbie is also featured in the Ben Mezrich book "The 37th Parallel" which is a chronological account of her brother Chuck Zukowski's UFO and cattle mutilation investigations. In 2017 she was awarded both the "Roswell Incident Excellence Award" by the Roswell Daily Record, and MUFON "Field Investigator of the Year." In 2021, Debbie released her first book, "The Alien Colonization of Earth's Waterways: A Reference Guide to UFO/USO Water Related Activity."

Debbie's full-time occupation is small family business owner-manager in Imperial, Missouri of an RV, and boat storage facility. Debbie became a PADI Scuba Dive Instructor in 2007 and holds teaching credentials in several dive instructor specialties including divers with special needs. She currently serves on the Board of Directors of "World of Water Foundation", a 501c3 dedicated to giving disabled veterans and children with special needs, the chance to experience scuba diving and snorkeling.

Contact email: mufondiver@gmail.com
Website: www.debbiediver.com

Margie Kay is a veteran 40-year paranormal investigator. She is the Owner and CEO of the UnX Network and the Un-X Media publishing company, serves as Assistant State Director for Missouri MUFON, is a Star Team Investigator, and is the Director of the OZ Interdimensional Communication Institute. She is a licensed private investigator in Missouri and owns and operates a fire investigation company. Margie has completed over 570 UFO investigations for MUFON and over 1,700 paranormal and UFO investigations on her own.

Kay is a self-taught psychic medium and remote viewer and has helped to solve over sixty homicides, theft, and missing person cases for law enforcement, private investigators and individuals in the U.S., South America, and Europe. She also assists other investigators with unsolved UFO and paranormal cases.

Margie is an accomplished classical string bass player and jazz/fusion electric bass player and worked as a professional musician for 20 years. She launched the KUNX Digital Broadcasting Network on October 31, 2021. There are multiple paranormal programs on the network.

Margie is the author of The Kansas City UFO Flaps; The Fast Movers: Evidence of High-Speed UFOs; Haunted Independence Missouri; Gateway to the Dead: A Ghost Hunter's Field Guide; The Remote Viewing Workbook; Winged Aliens; The Masters Dowser's Chart Book; A Sonoma County Phenomenon; and more. Some of Margie's experiences are chronicled in the books "*Paranormal Missouri: Show Me Your Monsters,*" and *"What Lurks Beyond: The Paranormal in Your Backyard"* by Jason Offutt, and in *"UFOs in Missouri"* by Lee Prosser. Kay also contributed a chapter to the book "A Greater Reality" by Rey Hernandez at his request.

Margie Kay has hosted radio shows since 2003. She has interviewed dozens of paranormal and UFO researchers and authors including Kathleen Marden, Stanton Friedman, Edwin Mitchell, Craig Campobasso, and many more. Margie's Un-X News Radio/Podcast on her YouTube Channel is broadcast periodically. She has appeared on numerous radio shows including Coast to Coast AM with Cheryll Jones and George Noory, Heidi Hollis, Jimmy Church, Jeff Rense, Joe Montaldo, and many others. She appeared on the pilot TV show Strange; on Hangar 1 UFO Hot Spot; on UFO Witness (The Cube) with Ben Hansen, and news programs on local channels and CNN.

Margie has written over 250 articles for magazines and news sites. Her articles have appeared in the MUFON Journal, Un-X News Magazine, Examiner.com, The Un-X Blog, and more.

Kay has spoken at over two hundred conferences and meetings nationwide including the Psychical Research Society, the Psychic Studies Institute, Missouri MUFON, Arkansas MUFON, PARACON, Mysteries of the Universe, Phenomecon, the Nebraska Bigfoot Conference, the Arkansas UFO and Paranormal Conference, The Midwest Conference on the Unknown, and is slated to speak at the Mega Con in Las Vegas. She has hosted and served as the MC for conferences since 1993.

Margie does remote viewing and dowsing workshops via ZOOM several times a year. Her students have had remarkable success using her unique methods.

Kay has been a UFO/alien experiencer since childhood. She has been in contact with these beings both in person and telepathically. They have healed her on several occasions. Margie is currently working on a book about these experiences and is planning to publish it in July of 2023.

Websites: www.margiekay.com www.unxnetwork.com www.unxmedia.com
Blog: un-xnews.blogspot.com
YouTube Channel UnX News Radio/Podcast https://www.youtube.com/channel
Facebook: www.facebook.com/margiekay11 Twitter: twitter.com/Margiekay06

UNXMEDIA
PUBLISHING

Publications by Un-X Media

Family Secrets 2017 by Jean Walker
Haunted Independence, Missouri 2017 by Margie Kay
Gateway to the Dead: A Ghost Hunter's Field Guide 2013 by Margie Kay
The Kansas City UFO Flaps 2017 by Margie Kay
The Remote Viewing Workbook 2019 by Margie Kay
A Sonoma County Phenomenon 2020 by Margie Kay
Un-X News Magazine 2001-2015 and 2001-2022
The Color Therapy Wall Chart 1999
The Fast Movers by Margie Kay, Bill Spicer, and Wayne Lawrence 2020
Doorway to Spirit by Devin Listrom 2020
The Alien Colonization of Earth's Waterways by Debbie Ziegelmeyer 2021
The Master Dowser's Chart Book by Margie Kay 2022 (lulu.com)
Mandala Meditation Adult Coloring Book by Margie Kay 2023
Dying to Meet Them: One Woman's Incredible Journey from NDE to UAP
by Mindy Tautfest 2023
The Unseens by Jeannette LaTulippe
50 Year Anniversary of the SE Missouri Ozarks UFO Flap
by Debbie Ziegelmeyer and Margie Kay 2023

And coming soon:

THOR: The Extraterrestrial on Earth by Margie Kay 2023

Documentary films are in the works for 2023

www.unxmedia.com

Email: editor@unxmedia.com

All books available at Amazon.com

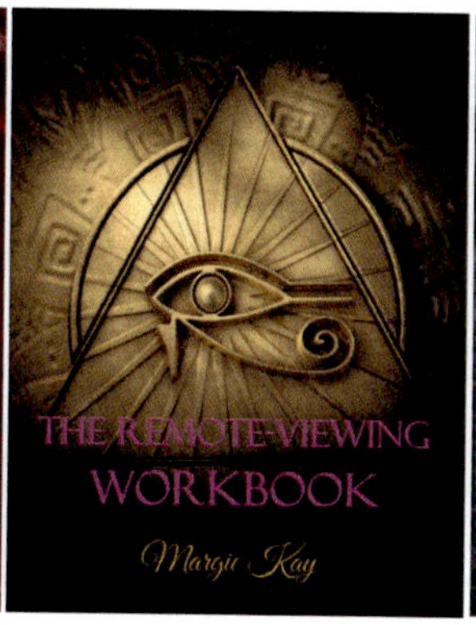

Made in United States
Orlando, FL
04 November 2026

90204702R00093